EXPERIMENTAL DESIGN

Other Griffin books on Experimental Design etc.

EXPERIMENTAL DESIGN

SELECTED PAPERS OF

FRANK YATES, C.B.E., F.R.S.

Formerly Head of Statistics Department
Rothamsted Experimental Station

GRIFFIN LONDON

CHARLES GRIFFIN & COMPANY LIMITED
42 DRURY LANE LONDON W.C.2

First published in 1970

SBN: 85264 168 0

PRINTED IN GREAT BRITAIN BY
COMPTON PRINTING LIMITED
LONDON AND AYLESBURY

FOREWORD

The complex organization of scientific research today may ensure that chance makes little contribution to progress. Forty years ago this was not so: the good fortune that caused R. A. Fisher (then reaching the peak of his powers) to meet a young mathematician from the Gold Coast Geodetic Survey, and to see the potential he had as an assistant and collaborator, had tremendous consequences for the development of statistical science. FRANK YATES came to Rothamsted Experimental Station in 1931 as assistant statistician. Two years later, Fisher left Rothamsted and Yates became head of the small but growing Department of Statistics.

Appreciation of the need for quantitative studies in agriculture, for which Yates himself had considerable responsibility, began a phase of development made still more rapid by the war. In 1947, Yates was instrumental in founding the Agricultural Research Statistical Service, by means of which Rothamsted expertise could be made available to the many small (and some large) research centres that still had no statisticians on their staffs. The appearance of the electronic computer, and Yates's pre-eminent role in adapting it to the needs of agricultural research statistics, increased still further the importance of his Department. In 1958, his appointment as a Deputy Director was a recognition that he participated to the full in all Rothamsted activities. Among many other honours that have come in recent years are the award of the Royal Society Medal in 1966 and the Presidencies of the British Computer Society (1960–61) and the Royal Statistical Society (1967–68). 'Retirement' in 1968 may have lessened his administrative responsibilities, but it has freed him to work even more intensively at his special concern of programming computers for statistical analysis.

Although officially associated for only two years, Ronald Fisher and Frank Yates became close friends and collaborators for 30 years. In many respects, the two have been complementary. From Fisher came the broad development of many fields of statistics, and notably the principles of the design and analysis of experiments. To Yates is mainly due the systematic elaboration of designs to meet a wide range of experimental situations. He gave the first clear account of the concepts of orthogonality and confounding, and first devised and used lattice, balanced incomplete block and quasi-Latin square designs. For factorial design and confounding, he has been a convincing advocate and a great innovator. He has also been largely responsible for sound development of the difficult subject of experiments on crop rotations.

His work has always been very close to the practical needs of the experimenter. Long before invention of the electronic computer, his experience in land survey had brought him a mastery of theory and numerical analysis involving least squares. Thence came his ideas on the processing of data from factorial experiments, on handling 'missing plots', and on the combination of inter- with intra-block information, all of which remain at the heart of modern computer techniques. He examined the influences of sampling errors within plots on the experimental variance, of course considering the economics as well as the mathematics of the sampling. He has always stressed the need for proper randomization in experiments, and has shown how this can be most conveniently achieved. We must indeed regret that he has never published a book on experimental design; his famous 'Technical Communication No. 35' has had tremendous influence, but a full exposition of his views on the 'Why' as much as on the 'How' of design would be most welcome.

This volume is intended to give a representative selection of Yates's publications on experimental design, with particular attention to some of the earlier papers now not readily accessible to statisticians. The reader will not find here a comprehensive textbook, but he can learn much about the practice of good design that is as true today as when first written. Although developed in connexion with agricultural research, Yates's ideas are equally relevant to any investigations in which comparative experiments are used, and his techniques have been adopted throughout scientific and technological research.

Those who have worked closely with Yates will have been impressed with two among his special skills, that of speedily seeing one or two anomalous entries in a large table (whether arithmetical faults or scientific 'leads') and that of writing down, as if intuitively, the orthogonal contrasts most relevant to the analysis of a body of data. The first can scarcely be illustrated in print, but the second is evident in several of these papers. Unquestionably this ability to dissect data, cultivated by his research, has given Yates exceptional facility in statistical analysis; it lends strength to his view that a thorough grounding in the practice of experimental design and analysis should be a major component in the training of a biometrician. Yet his concern has always been as much with the experiment as with the design; his interest has only stopped short of generalizations, extensions, and combinatorial legerdemain for which he could see no application. Like Fisher, he has always emphasized the gain to understanding that comes from doing the arithmetic of an analysis and not seeking to learn solely from the algebraic theory.

Unfortunately this volume cannot find space for examples of Yates's publications in other fields. Paralleling his work on design, he was among the first to build up the second major pattern of modern data collection, sample survey theory and practice. Many papers and a widely-acclaimed book (now in its third edition) have established methods of general applicability that are now standard. His wartime operational research on crop responses to fertilizers and their relation to agricultural policy illustrates well his skill in welding the accumulated results of research into a sound basis for action: one need not hold any purely utilitarian view of research to appreciate how, here and elsewhere, Yates persuades findings from diverse sources to tell a coherent and practical tale.

Since 1954, his efforts have been directed primarily at computer programming, with special reference to experimentation and sample survey. Though less productive of published papers, this phase of his career has proved important and influential. Characteristically, he has not been content merely to translate earlier routines of analysis into a computer language. He has sought to build into programs his own experience in the interpretation of data and the detection of anomalies, he has been concerned to guard against accumulation of numerical inaccuracy, and he has planned for flexibility of input and output that gives the user of a program a wide choice while preserving the essential features of sound analysis. Moreover he has continually aimed at useful generality, in order that most standard designs (with the modifications arising because of covariance analysis, missing data, transformations, and so on) can be handled within one program package.

We first knew Frank Yates at various periods as our master. We write now to mark the official retirement of a friend, one to whom we all owe much. From him we learned something of the discipline of scientific research, of exactness of expression in word as well as number, and of the close interaction with non-statistical colleagues so essential if a statistician is to play his proper part as a collaborator in research. From him we gained a generous flow of ideas that stimulated our own early research, as well as many kindnesses that have helped our careers. In expressing our gratitude, we present a volume that can show a new generation one major aspect of the work of a great scientist: we hope the reader may also glimpse, in an incisive phrase, a neat turn of logic, or a progression from intuition to logical proof, the man inside that scientist's skin.

July 1969

W. G. Cochran
D. J. Finney
M. J. R. Healy

AUTHOR'S PREFACE

When Cochran, Finney and Healy told me they would like to mark my retirement by publishing, in book form, a collection of some of my more important papers on the lines of Fisher's 'Contributions to Mathematical Statistics', I was both surprised and pleased, but somewhat dubious whether the material was of sufficient continuing interest to justify republication. After discussion, however, we agreed that a volume containing papers on a single topic might overcome one of the main objections to 'Collected works'.

A collection of papers on experimental design also attracted me because this branch of statistics, which is of very great practical use, has been sadly misunderstood by many statisticians, particularly teachers of statistics who are not themselves faced with the task of actively designing experiments. Surprisingly many elementary statistical textbooks treat the subject in a most cursory and erroneous manner, often expounding the analysis of variance before mentioning experimental design, and failing entirely to recognize that the validity of the analysis of variance depends on the adoption of the appropriate randomization procedure. Nor are all advanced textbooks above reproach.

This is particularly regrettable because the elements of experimental design and analysis form a good introduction to any general statistical course. The basic ideas and arithmetical operations required for analysis are simple. Moreover it is only from properly designed experiments that rigorous statistical inferences can be drawn; observational material is always subject to qualifications and doubts regarding its interpretation. It is important that students should get this fundamental difference clear in their minds.

The present series of papers covers an exciting period in the development of designs suitable for experiments on agricultural crops grown in the field. Other applications need somewhat different designs which in part depend on the types of variability that are encountered, but the principles are the same. Mathematicians studying the papers must not expect advanced or complicated mathematics. In experimental work dealing with quantitative variates all that is required is intelligent application of classical least squares theory. In recent years the theory of this branch of statistics has been extended by the development of matrix algebra, but this is not used in the papers reproduced here as its use only became popular after they were written. This, I think, is no great disadvantage for the student as he may well be able to perceive better the structure of an experiment if it is not concealed by the shorthand of matrix algebra.

Although Fisher originally resolved the problem of computing the error of randomized block experiments by the application of least squares, this approach was largely lost sight of subsequently and attempts—not always successful, as will be seen by a study of Paper I—were made to construct appropriate methods of analysis for particular experimental designs by rule-of-thumb applications of the analysis of variance technique. It was fortunate that when I first took up statistics at Rothamsted I had had no formal training in the then fashionable branches of the subject, but had acquired considerable familiarity with classical least squares in the course of geodetic survey work. This, I think, is one of the reasons why I found the subject of experimental design and analysis particularly easy, and why others found it difficult.

I was largely guided by the sponsors in the selection of papers for inclusion. Permission by the editors and publishers of the various journals to reproduce these papers is gratefully acknowledged. At the sponsors' suggestion I have also included a full bibliography of my papers, with brief notes on the contents of papers on experimental design not included in this volume.

I have added a brief note at the beginning of each paper to put it in historical perspective, but for the most part I leave the papers to speak for themselves. As the reader will see, they have a severely practical orientation, as is to be expected when the subject is being developed for practical use in actual research rather than as an academic exercise. Methods of computation in particular are fully dealt with. The student is advised to follow these computations through on a desk calculator, if he wishes to obtain a thorough understanding of what is being discussed. Many of the detailed methods of computation, however, are no longer very relevant when an electronic computer, properly programmed for the analysis of experiments, is available. Missing plots, for example, can now be handled by a general and very simple iterative procedure which is particularly suited to computers.

In recent years there have been various new accretions to experimental design, not all of them to my mind useful; some of them are briefly discussed in Paper XII. I would particularly stress the danger, when devising new designs, of overlooking commonsense points through over-simplified theory. I myself have not been guiltless in this respect, as witness the oversight on quasi-Latin squares which is mentioned in Paper XII.

May, 1969 FRANK YATES

CONTENTS

Paper		Page
I	The principles of orthogonality and confounding in replicated experiments	1
II	The analysis of replicated experiments when the field results are incomplete	41
III	The formation of Latin squares for use in field experiments	57
IV	Complex experiments	69
V	The analysis of groups of experiments	119
VI	A new method of arranging variety trials involving a large number of varieties	147
VII	Incomplete randomized blocks	181
VIII	The recovery of inter-block information in balanced incomplete block designs	201
IX	Analysis of data from all possible reciprocal crosses between a set of parental lines	211
X	The analysis of experiments containing different crop rotations	229
XI	Principles governing the amount of experimentation in developmental work	255
XII	A fresh look at the basic principles of the design and analysis of experiments	265
Note on fixed and random effects models		281
Bibliography		287
Index		295

PAPER I

THE PRINCIPLES OF ORTHOGONALITY AND CONFOUNDING IN REPLICATED EXPERIMENTS

FROM THE JOURNAL OF AGRICULTURAL SCIENCE
VOLUME XXIII, PART I, pp. 108–145, 1933

Author's Note

In the 1920s Fisher laid the foundations of modern experimental design. The initial step was the evolution of methods to estimate error in commonly used designs, such as arrangements in blocks. A further important step was the arrangement of the computations in the form of the analysis of variance. Randomization was introduced to ensure the validity of the exact tests of significance made possible by Student's t-distribution and Fisher's z-distribution. Equally important was Fisher's emphasis on the value of factorial design.

These methods were used for the analysis of increasingly complex agricultural field trials laid out at Rothamsted and elsewhere. The complexities were introduced mainly because of the desire of the experimenters to get more out of the results of a single experiment. In particular, the introduction of variants of the main treatments led to confounded designs, without any very clear understanding of the effect this had on the analysis of variance procedure.

This tangle is resolved in Paper I. For some time I had had a feeling of disquiet at Wishart's *Arch. für Pflanzenbau* paper (reference in Paper I), without being able to put my finger on what was wrong. Suspicion hardened to certainty when I obtained what I regarded as an impossibly low error—2% per plot—from the analysis of a smaller experiment of the same type, but with supplementary treatments on columns only. This set me furiously to think, and because the experiment was simpler, and because I had done the computations myself, the thinking bore fruit. Once the property of orthogonality was formally recognized the road was clear for the confident use of the analysis of variance and for the further development of complex designs.

1969 FRANK YATES

THE PRINCIPLES OF ORTHOGONALITY AND CONFOUNDING IN REPLICATED EXPERIMENTS.

By F. YATES, B.A.

(*Rothamsted Experimental Station, Harpenden.*)

(With Seven Text-figures.)

I. Introduction.

In the past few years the procedure of the analysis of variance, developed by Dr R. A. Fisher, has been widely employed in agricultural experimental work, in conjunction with suitably designed field trials, and in consequence has been used more and more by experimenters who have little statistical training. The actual computations necessary for the analysis of a well-designed experiment are extremely simple, and even the more complex experiments can be dealt with by repeated applications of what is essentially one process. This simplicity depends on the fact that the experiments are designed so as to be orthogonal. In cases of non-orthogonality the analytical procedure must be considerably modified.

Orthogonality is that property of the design which ensures that the different classes of effects to which the experimental material is subject shall be capable of direct and separate estimation without any entanglement. In a randomised block experiment, for instance, in which every treatment occurs once in every block, the fertility differences between the different blocks can be estimated from the differences of the means of all the plots within each block, since an increase in the effect of any one treatment will increase each block mean equally (at least in so far as the treatment effect is independent of fertility level); similarly the treatment effects may be estimated by the differences of treatment means, which are unaffected by differences in fertility in different blocks. In such an experiment blocks and treatments are said to be orthogonal with each other. If, however, each block contains plots receiving some of the treatments only, comparisons between treatments will be affected by differences of fertility between the blocks, and similarly comparisons between blocks will be affected by differences between the effects of the treatments. The blocks and treatments are then no longer orthogonal.

Lack of orthogonality may be due to accident, as when some of the plots of a field trial suffer from damage, so that their yields are virtually unknown; it may be due to some error of design; or it may be unavoidable owing to the nature of the material, as for instance in livestock trials on animals of which the sex cannot be determined at the commencement of the experiment, for then sex and treatments will be non-orthogonal. In all these cases non-orthogonality is a disadvantage, complicating the analysis, and lowering the efficiency of the experiment.

The first part of this paper contains an account of the methods of analysis appropriate to the various types of non-orthogonality, except that caused by a few missing values, which can be more appropriately discussed separately. The experiment on sugar beet carried out at the Rothamsted Experimental Station in the year 1929, which was in some respects non-orthogonal, is taken as an example; the original analysis has been described in a paper by J. Wishart (1), the experiment being treated as if it were orthogonal, and the results and methods there given therefore require correction.

Non-orthogonality is sometimes deliberately introduced into experimental design as a consequence of what is termed confounding. If, in order to keep the number of plots in each block of a complex experiment small, complete replication within each block is sacrificed, certain treatment differences will be entangled with fertility differences between blocks, so that treatments and blocks are non-orthogonal as far as these differences are concerned. Descriptions of experiments involving this principle have already been published (2, 3), but no very explicit account of the underlying theory has been given. Methods have recently been devised which put the analysis of confounded experiments on a more satisfactory basis, and these should do much to remove the distrust which has sometimes been felt at its use. There is no doubt that confounding considerably widens the scope of experiments of the complex type, which embody questions on several different points. Such experiments are by far the most efficient, and in addition give conclusions which, being founded on a wider inductive basis, are of more certain validity (2, 4). Any method which enables this type of experiment to be more widely employed is therefore likely to be of great practical utility.

In confounded experiments, where the non-orthogonality is deliberate, the design should be such that the ordinary methods of analysis require very little modification, a slight rearrangement of the data enabling the orthogonality which was apparently lost to be regained. The discussion of confounding given in this paper does not pretend to be exhaustive, only the general principles underlying the simpler types of confounding being dealt with. There is wide scope for the design of new experimental patterns, and the complete enumeration of the more simple types would be well worth while. On the other hand there is no need to await such enumeration before making use of the principle, and the present discussion should enable the experimenter to devise designs to suit his own particular needs.

II. General discussion of non-orthogonality.

In an ordinary replicated field experiment of the randomised block or Latin square type the differences of the means of plots receiving the same treatments are taken without hesitation to be true measures of treatment differences, but this is only so because the experiment has been specially arranged so as to be orthogonal. What exactly happens in cases of non-orthogonality may best be illustrated by a concrete example. Suppose that it is desired to determine the effect of two treatments on the growth of chickens, these treatments being applied each to one-half of a batch of chickens. (The chickens which are to receive the first treatment must of course be chosen at random and be kept in the same pen as the others.) In the absence of a method determining the sex at hatching the proportion of cockerels in the two groups will not be the same, and since cockerels grow faster than pullets the treatment which happens to have been given to the group containing the greater number of cockerels will clearly be given an unfair advantage if the average final weight of each group is taken as a measure of treatment effect.

In non-orthogonal experiments the ordinary methods of the analysis of variance require modification. It is true that if sex were neglected in the experiment described above differences in the mean effect of treatments which were in reality due to sex would not in general be judged significant in an ordinary analysis, since the estimate of error would be increased, but it is clear that more accurate conclusions can be drawn if the effect of sex is eliminated.

The methods of the analysis of variance applicable to a classification with unequal numbers in the different classes are well known ((5), § 44). The variance between classes and the variance within classes can be simply calculated and compared by means of the z test. The variance within classes furnishes a valid estimate of "error" in the sense ordinarily implied by that term. If, however, the various classes (here called sub-classes) can themselves be arranged in a multiple classification[1] new problems arise. Analogous difficulties are encountered in cases where the error is estimated from some interaction of an incomplete multiple classification.

A general method of analysis, which is applicable to all experiments with multiple classifications, is provided by the fitting of constants by the method of least squares. Tests of significance are made by fitting constants to represent all effects other than the one to be tested, evaluating the residual variance between classes after fitting has been performed, and comparing this variance with the intraclass variance. For example, in order to test whether any interaction between treatments and sex exists in a poultry experiment of the type already described, constants representing treatments and sex can be fitted, and the residual variance between classes after this fitting can be tested for significance by means of the z test.

This test of significance presupposes that there is an intraclass variance with which to compare the residual variance. When the error is estimated from some interaction, there being only a single member of each sub-class, and the multiple classification is incomplete, the fitting of constants also provides a method of obtaining the error interaction variance. Agricultural replicated field trials which have one or more plot yields missing, or are subject to some defect of design, so that orthogonality is not secured, may be dealt with by this method.

The fitting of constants serves a further purpose. It is frequently reasonable to suppose that the interactions between the main effects are negligible. If this is so (the experiment itself will furnish evidence on the point) the most efficient estimates of the magnitudes of the main effects may be made by fitting constants to represent these effects only. The significance test for any set of effects estimated on this assumption can be made by the method of residual variance described above.

If the interactions cannot be ignored the efficient estimates of the main effects are the means of the sub-class means (assuming the multiple classification to be complete). In orthogonal experiments these estimates are precisely the same as those obtained on fitting constants to represent the main effects only. In orthogonal experiments, therefore, there is no

[1] The term *multiple* (*double, triple, etc.*) *classification* is used to denote a simultaneous classification of several different sets of classes; the term *n-fold classification* is used to denote an ordinary classification containing n classes, which may themselves form one set of classes from a multiple classification. Thus a 5×5 Latin square would be described as a triple five-fold classification, with one member in each sub-class.

need to consider whether the interactions are in fact negligible, when estimating the main effects. In experiments where some of the sub-classes are missing entirely we can make no estimate at all of the average main effects unless we assume that some at least of the interactions are negligible. This is an extreme case, and as orthogonality is approached more and more nearly the estimates based on the two different assumptions become closer and closer. It is important to notice, however, that there are two separate tests of significance based on these different assumptions in all non-orthogonal experiments. On the other hand, since it is logically impossible that an interaction should exist without a main effect, the significance of main effects should strictly be tested on the assumption that their interactions are negligible, no test at all being necessary if the interactions are significant. The tests based on the assumption that the interactions exist are, however, simpler, and may be used with safety in cases which approach orthogonality.

Although the fitting of constants always provides efficient estimates and tests of significance the process is one which involves lengthy calculations, especially in the more complex types of experiment, and it is therefore important to utilise shorter methods when these are available. The following is a list of tests which can be made on non-orthogonal data without fitting constants in cases where an estimate of intraclass variance is available.

(1) Any set of main effects may be tested on the assumption that all other main effects and all interactions exist and are not negligible.

(2) The same test may be applied to any set of interactions of which the variance can in orthogonal cases be calculated from the sum of the squares of the deviations of a set of numbers from their mean (*e.g.* the interactions of a $2 \times n$ table, which can be calculated from the differences of the two rows of entries).

(3) An experiment in which each classification is two-fold only (*i.e.* of which the results can be arranged in a $2 \times 2 \times 2 \times \ldots$ table) can be completely analysed whether interactions are assumed to exist or not. The same methods are applicable to experiments which are only partially two-fold.

(4) A useful approximate method is available which is applicable to all cases except those where some of the sub-classes of a multiple classification are missing entirely. The method can be extended to cases where only a few sub-classes are missing by means of the missing plot technique (6) which can, by a process to be described in a later paper, be usefully employed even when more than one plot is missing.

The procedure of fitting constants and the shorter methods of this list are illustrated in the next section.

Although when orthogonality is not attained the procedure of the analysis of variance appropriate to orthogonal experiments breaks down its failure is not always immediately apparent, and consequently serious errors may be introduced. It is therefore important that even experimenters who ordinarily have to deal only with orthogonal data should be able to recognise the cases of non-orthogonality which may occasionally occur through some defect in design or through some accident. Certainly anyone who wishes to employ the methods of confounding, discussed in the later sections of this paper, in the design of experiments must have clear ideas on orthogonality, or he may find himself faced with the analysis of experiments involving lengthy and cumbersome calculations.

III. METHODS OF ANALYSIS.

Fitting constants.

The method of fitting constants (7, 8) may best be illustrated in its application to a concrete example. Table I shows results such as might be obtained in a poultry experiment of the type already described, but with three treatments (the numbers of birds in each class being shown in brackets).

Table I. *Mean weight per bird.*

Treatment	*A*	*B*	*C*
Cockerels	2·82 (5)	2·50 (9)	2·75 (12)
Pullets	1·99 (10)	1·83 (6)	1·82 (3)
Mean	2·405	2·165	2·285

The estimate of the variance within classes, σ^2, which can be computed from the deviations of the individual bird weights from the subclass means shown in the table, is taken as 0·0786. This will be based on 39 degrees of freedom.

In order to test the significance of the interactions between treatments and sex, constants representing the direct treatment and sex effects must be fitted. The procedure is as follows. Taking constants t_1, t_2, t_3 to represent the mean weights of equal numbers of cockerels and pullets subjected to the three treatments, and $2c$ to represent the mean difference between cockerels and pullets, w being the weight of an individual bird, the values of the constants must be so chosen as to minimise

$$S\{w - (\pm c + t_s)\}^2,$$

where the summation is taken over all birds, the signs of c and the particular t being chosen according to the scheme of Table II.

Table II.

Treatment	*A*	*B*	*C*
Cockerels	t_1+c	t_2+c	t_3+c
Pullets	t_1-c	t_2-c	t_3-c

The equations for determining the values of the t and c can be written down by the method of least squares. Reference may be made to (5), § 29, where the application of the method to partial regressions is given. The method here is formally the same: c, t_1, t_2, t_3 correspond to the regression coefficients b, w to y, and the equations

$$W = \pm c + t_s$$

to the regression equation

$$Y = b_1 x_1 + b_2 x_2 + b_3 x_3 + b_4 x_4,$$

the values of all x being limited to ± 1 and 0. Thus the equations for determining the t and c are obtained by forming the set of equations

$$b_1 S(x_1^2) + b_2 S(x_1 x_2) + b_3 S(x_1 x_3) + b_4 S(x_1 x_4) = S(x_1 y),$$

etc., with the special values of x and y relevant to this particular experiment, the summation being taken over each individual bird. For this purpose it is useful to make out a table of the total weight of all birds in each class (Table III).

Table III. *Total bird weight.*

Treatment	*A*	*B*	*C*	Total
Cockerels	14·10 (5)	22·50 (9)	33·00 (12)	69·60 (26)
Pullets	19·90 (10)	10·98 (6)	5·46 (3)	36·34 (19)
Total	34·00 (15)	33·48 (15)	38·46 (15)	105·94 (45)

The equations can now be immediately written down. They are

$$\begin{aligned}
45c \quad - 5t_1 \quad + 3t_2 \quad + 9t_3 &= 33{\cdot}26,\\
-5c \quad + 15t_1 \qquad\qquad\qquad &= 34{\cdot}00,\\
+3c \qquad\qquad + 15t_2 \qquad\quad &= 33{\cdot}48,\\
+9c \qquad\qquad\qquad\quad + 15t_3 &= 38{\cdot}46.
\end{aligned}$$

The numerical solution is

$$c = 0{\cdot}3970, \quad t_1 = 2{\cdot}3990, \quad t_2 = 2{\cdot}1526, \quad t_3 = 2{\cdot}3258.$$

We are now in a position to determine whether the assumption of negligible interaction is in fact justified. The fitting of the constants t and c will account for 3 degrees of freedom (after allowance has been made for the mean). The reduction in the total sum of squares due to the fitting, given by the general formula

$$S(y^2) - S(y - Y)^2 = b_1 S(x_1 y) + b_2 S(x_2 y) + b_3 S(x_3 y) + b_4 S(x_4 y),$$

is therefore

$$33{\cdot}26 \times 0{\cdot}3970 + 34{\cdot}00 \times 2{\cdot}3990 + \text{etc.} = 256{\cdot}2895.$$

This includes the correction for the mean, 249·4064, and therefore the additional reduction in the sum of squares above that obtained by fitting the mean only is 6·8831. The total sum of squares between classes is 6·9872, and the analysis of variance may therefore be set out as in Table IV.

Table IV.

	Degrees of freedom	Sum of squares	Mean square
Constants c and t	3	6·8831	2·2944
Remainder (interaction)	2	0·1041	0·0520
Between classes	5	6·9872	
Within classes	39	3·0654	0·0786
Total	44	10·0526	

If the remainder of the sum of squares between classes (2 degrees of freedom) gives a variance significantly above that within classes it is evidence that the hypothesis that the effects of sex and treatments are additive is not justified, or in other words that interaction between sex and treatments exists. In this experiment there is clearly no evidence of interaction.

The significance of the direct effects of treatments may now be tested on the assumption of negligible interaction. This is done in a similar manner to the test for interaction, the residual variance after fitting a constant for sex only being obtained from the 3 degrees of freedom for sex and treatments. The sum of squares due to fitting the constant c only, 6·4128, can be immediately calculated from the total column of Table III, being the sum of squares between classes when treatments are ignored. The analysis is therefore:

	Degrees of freedom	Sum of squares	Mean square	z
Sex, neglecting treatments	1	6·4128	6·4128	
Remainder (treatments)	2	0·4703	0·2352	0·5480
Sex and treatments	3	6·8831		

The 5 per cent. point of z is 0·5876, and the effect of treatments as a whole cannot therefore be regarded as unquestionably significant.

In a similar manner the test for the effect of sex may be thrown in the analysis of variance form. The analysis will be:

	Degrees of freedom	Sum of squares	Mean squares
Treatments, neglecting sex	2	0·9992	0·4996
Remainder (sex)	1	5·8839	5·8839
Sex and treatments	3	6·8831	

It will be seen that the variance ascribable to treatments, neglecting sex, is considerably higher than the treatment variance. No relevant test of significance can be made on this variance, however, since it contains sex as well as treatment effects. The total bird weight for treatment C, 38·46, for example, is too high because of the large proportion of cockerels in this group.

The level of significance of sex and treatments, tested separately, may either be higher or lower than that of the combined effect. In extreme cases it may happen that although the variance corresponding to the 3 degrees of freedom for the constants c and t is significant, implying that either sex or treatments or both produce an effect, neither sex nor treatments, when tested separately, appears significant. The reason for this may easily be seen if we consider the case of an experiment with two treatments, where all the birds receiving treatment A are cockerels and all those receiving treatment B are pullets; it is then clearly impossible to isolate the effects of sex and treatment, however accurate the experiment.

The test for the significance of the direct effects of treatments when the interaction is not assumed to be negligible might be made on similar lines to the test for the significance of interactions, constants representing sex and the interactions being fitted. A shorter and more convenient procedure for this test is given below.

It is of interest to see how the method of fitting constants reduces to the ordinary recognised method of analysis when the data is orthogonal. The essential point of orthogonal data is that the numbers in the various classes are such that variation of one set of effects does not alter the totals representing any other set of effects. Mathematically this implies that all the product sums, $S(x_1 x_2)$, etc., of the equations for determining the fitted constants are zero. If this is so each equation contains only one constant, and can be solved immediately, the value so obtained being the same whether or not all the other constants are fitted at the same time. The reduction in the sum of squares due to fitting reduces to

$$\frac{\{S(x_1 y)\}^2}{S(x_1^2)} + \frac{\{S(x_2 y)\}^2}{S(x_2^2)} + \ldots,$$

from which it follows that the reduction in the sum of squares due to the fitting of one set of constants is the same whether other sets are fitted or not. From this the whole procedure of the analysis of variance as applied to orthogonal experiments can be immediately deduced.

In agricultural field experiments there is not usually replication within each class; in an ordinary randomised block experiment, for instance, each treatment occurs once only within each block, and the estimate of error variance is obtained by the interaction of blocks and treatments. Such experiments are ordinarily designed so as to be orthogonal, but if through some cause, either intended or accidental, non-orthogonality is introduced, the methods of analysis must be suitably modified, or serious errors will result. In some cases the missing plot technique is applicable, while in others (more particularly well-designed confounded experiments) the analysis can be so modified as to create virtual orthogonality, but in certain cases the only suitable method of full analysis is that of fitting constants. An example of this type is given later in the paper. In any event it is important to notice that an unjustified assumption of orthogonality will lead not only to false treatment variances, but also to a false error variance.

It may perhaps be as well to recall here that it is possible to test any linear function of the fitted constants by means of the t test. In R. A. Fisher's notation (5), § 29, if the solution of the set of equations for c and t be $c_{11}, c_{12}, c_{13}, c_{14}$, when 1, 0, 0, 0, are substituted for the numerical terms of the four equations, and $c_{21}, c_{22}, c_{23}, c_{24}$, when 0, 1, 0, 0, are substituted for the numerical terms, etc., then the variance of any linear function of c and t,

$$x = l_1 c + l_2 t_1 + l_3 t_2 + l_4 t_3,$$

is given by

$$V(x) = \{l_1^2 c_{11} + l_2^2 c_{22} + \ldots + 2 l_1 l_2 c_{12} + \ldots\} \sigma^2,$$

it being noted that $c_{12} = c_{21}$, etc.

In our example the values of c_{11}, etc., are

225	75	− 45	− 135
75	585	− 15	− 45
− 45	− 15	569	27
− 135	− 45	27	641

all divided by 8400. These values can, of course, be used to determine the values of the constants themselves.

The estimate of the difference of treatments A and B, for example, is $t_1 - t_2$, or 0·2464. The variance of this quantity is given by

$$V(t_1 - t_2) = \sigma^2 (1 \times 585 + 1 \times 569 + 2 \times 15)/8400 = 0{\cdot}1410\sigma^2.$$

The estimate of the standard deviation is therefore 0·1053 (39 degrees of freedom) and t is in the neighbourhood of the 1 in 50 point. This, however, is the greatest of the three differences, and unless there is *a priori* justification for singling out this difference for special examination the test cannot be considered valid.

Variance from weighted squares of means.

Instead of fitting constants it is possible to perform a direct test for treatments on the means of Table I. The variances of the three treatment means are $\frac{1}{4}(\frac{1}{5} + \frac{1}{10})\sigma^2$, $\frac{1}{4}(\frac{1}{9} + \frac{1}{6})\sigma^2$, and $\frac{1}{4}(\frac{1}{12} + \frac{1}{3})\sigma^2$, or $\sigma^2/13{\cdot}3333$, $\sigma^2/14{\cdot}4$, and $\sigma^2/9{\cdot}6$. The efficient estimate of σ^2 from these means is therefore

$$\frac{1}{2}\left[2{\cdot}405^2 \times 13{\cdot}3333 + 2{\cdot}165^2 \times 14{\cdot}4 + 2{\cdot}285^2 \times 9{\cdot}6 - \frac{(2{\cdot}405 \times 13{\cdot}3333 + 2{\cdot}165 \times 14{\cdot}4 + 2{\cdot}285 \times 9{\cdot}6)^2}{13{\cdot}3333 + 14{\cdot}4 + 9{\cdot}6}\right],$$

as can be seen on the analogy of interclass variance with unequal numbers in the various classes. The numerical value of this variance is 0·1995, as compared with the value 0·2352 previously obtained, but the two tests are on a different basis, only the first being made on the assumption of negligible interaction. It can be shown that the present test is efficient if interaction is not to be neglected, the variance obtained being identical with the residual variance when constants representing sex effect and interaction are fitted. That this is so is clear on general grounds, since if interaction exists the three treatment means contain all the information possible on average treatment effect.

This suggests an alternative method of testing the significance of the interactions, which may be estimated from the differences of mean bird weight between cockerels and pullets for treatments A, B and C. These differences are 0·83, 0·67 and 0·93 respectively, and their variances are four times the variances of the corresponding treatment means. The estimate of σ^2 can be made in exactly the same manner as was employed in the case of the treatment means, the value obtained, 0·052032, agreeing precisely with that obtained by the method of fitting constants. This method of computation is obviously preferable to the method of fitting constants, but it is only applicable to interactions when these are capable of direct calculation from differences, as is the case in any classification of the $2 \times 2 \times 2 \times \ldots \times n$ type.

The method of weighted squares provides a useful test for the main effects in cases where these effects may appropriately be estimated by the direct means of the sub-class means, that is in cases which approach orthogonality or where the assumption of negligible interaction is not justified. With negligible interactions and widely differing numbers in the different classes, and particularly when some of the classes are missing entirely, a considerable amount of information may be lost, and in such cases the fitting of constants should be resorted to, at any rate if the data justify the extra labour.

The actual amount of information lost, on the assumption of negligible interaction, by the alternative method, can only be determined in any particular experiment after the values of c_{11}, etc., have been found. If the intraclass variance is denoted by B, and the variance ascribable to treatment effects by A ((5), § 40), it is possible to evaluate the expectation, in terms of A and B, of the treatment variances obtained by the two methods of analysis. In the particular example given that of the method of fitting constants is $13{\cdot}8765\ A + B$, while that of the other method is $12{\cdot}2742\ A + B$. Thus 89 per cent. of the information is utilised by the second method.

Multiple two-fold classifications.

When interactions are assumed to exist experiments containing only two-fold classifications may clearly be completely analysed by the method of weighted squares of means; in this type of classification the method is equivalent to the simpler approximate method described below. The direct effects may also be estimated and tested for significance on the assumption of negligible interaction, by a method now to be described. The same method may be applied to the two-fold parts of experiments not wholly two-fold, and the procedure may therefore be illustrated by applying it to the sex difference in the example already given.

If the interaction be negligible the best estimate of the difference between cockerels and pullets will be obtained by taking the weighted mean of the differences for treatments A, B and C. These differences are 0·83, 0·67 and 0·93, their variances being $(\frac{1}{5} + \frac{1}{10})\,\sigma^2$, $(\frac{1}{9} + \frac{1}{6})\,\sigma^2$ and $(\frac{1}{12} + \frac{1}{3})\,\sigma^2$. Weighting inversely as the variances, the weighted mean 0·7940 is obtained. Its variance will be the reciprocal of the sum of the reciprocals of the above variances, or $0{\cdot}107143\sigma^2$. The significance can be tested in doubtful cases by the t test. It should be noticed that the above estimate is precisely equal to the estimate, $2c$, obtained by fitting constants, and that its variance is equal to the variance of $2c$ calculated from c_{11}.

When the interaction sum of squares is calculated by the method of weighted squares of means, instead of by fitting constants, the differences of t_1, t_2 and t_3 will not be available as estimates of the treatment differences. Instead the weighted means of the differences for cockerels and pullets may be employed. The differences between treatments A and B, for instance, are 0·32 and 0·16 respectively, with variances $(\frac{1}{5} + \frac{1}{9})\sigma^2$ and $(\frac{1}{10} + \frac{1}{6})\sigma^2$. Weighting inversely as these variances the weighted mean 0·2338 is obtained, with variance $0{\cdot}1436\sigma^2$. This should be compared with the efficient estimate already obtained from the difference of t_1 and t_2. If instead of the weighted mean of the differences the unweighted mean, 0·24, had been taken, on the assumption of negligible interaction a slightly less efficient estimate would have been obtained. If, however, the interaction is not to be neglected the unweighted mean provides the efficient estimate of the treatment differences. The values of the three sets of estimates, their variances, and their efficiency on the assumption of negligible interaction, is given in Table V. In this experiment there is little loss of accuracy in using the weighted means, and very little additional loss in using the unweighted means, except for C–B.

Table V.

	Estimates			Variances/σ^2			Efficiency	
	Efficient	Weighted means	Un-weighted means	Efficient	Weighted means	Un-weighted means	Weighted means %	Un-weighted means %
A–B	0·2464	0·2338	0·24	0·1410	0·1436	0·1444	98	98
A–C	0·0732	0·1095	0·12	0·1563	0·1713	0·1792	91	87
C–B	0·1732	0·1828	0·12	0·1376	0·1402	0·1736	98	79

Approximate analysis of variance.

It is possible to perform an ordinary analysis of variance on the table of mean bird weights (Table II) on the assumption that each mean bird weight has an equal variance, the error variance being taken as the mean of the variances of the various mean bird weights, or

$$\tfrac{1}{6}\left(\tfrac{1}{5} + \tfrac{1}{10} + \tfrac{1}{9} + \tfrac{1}{6} + \tfrac{1}{12} + \tfrac{1}{3}\right)\sigma^2,$$

of which the estimate from the intraclass variance is 0·01302. The analysis now reduces to the ordinary orthogonal form, provided there are no missing classes, the values obtained being given in Table VI.

Table VI.

	Degrees of freedom	Sum of squares	Mean square
Treatments	2	0·05760	0·02880
Sex	1	0·98415	0·98415
Interaction	2	0·01720	0·00860
Intraclass	39	—	0·01302

Although the z distribution will not hold exactly it does not seem likely that any great disturbance will be introduced. Apart from this the method is equivalent to throwing away the information on the distribution of the class numbers amongst the various classes.

The chief utility of this approximate method lies in the testing of interactions of complex experiments which could otherwise be tested only by fitting constants. In the analysis of multiple two-fold classifications it is equivalent to the analysis by the method of weighted squares of means, and therefore provides a rigorous method when interactions are not negligible. (This equivalence, which depends on a simple algebraic identity, must follow from general considerations, and need not be demonstrated here.)

IV. THE CONFOUNDING OF MAIN EFFECTS.

In agricultural varietal trials it is frequently impracticable, owing to sowing difficulties, to use such small plots as are possible in manurial experiments. It is, moreover, often desirable to introduce different varieties into manurial trials, and *vice versa*, in order to give a wider inductive basis to any conclusions that may be drawn. This has led to the simple expedient of sowing large plots with each variety and subdividing these plots into smaller ones for the purposes of manurial comparisons.

The simplest type of lay-out of this nature is that of complete replication of the manurial treatments within each plot of the varietal treatments. An extended nomenclature will be necessary. We will call the varietal treatments of the above example the *main treatments*, the varietal plots the *main plots*, the manurial treatments the *sub-treatments*, and the manurial plots the *sub-plots*. Fig. 1 gives the plan of an experiment of this type, with the main plots arranged in randomised blocks.

The analysis of variance presents no great difficulties. Since every sub-treatment occurs once in each main plot the differences between main plots will properly represent the average main treatment effects for sub-treatments a, b and c, and the differences in different blocks will form an estimate of error. The first part of the analysis of variance is therefore as follows:

	Degrees of freedom
Blocks	3
Main treatments	3
Main plot error...	9
Total, main plots	15

The experiment may also be regarded as one of sixteen randomised blocks with three plots per block, the main plots corresponding to the blocks and the sub-plots to the plots. There is, however, this difference: the error sum of squares, which is derived from interaction between main

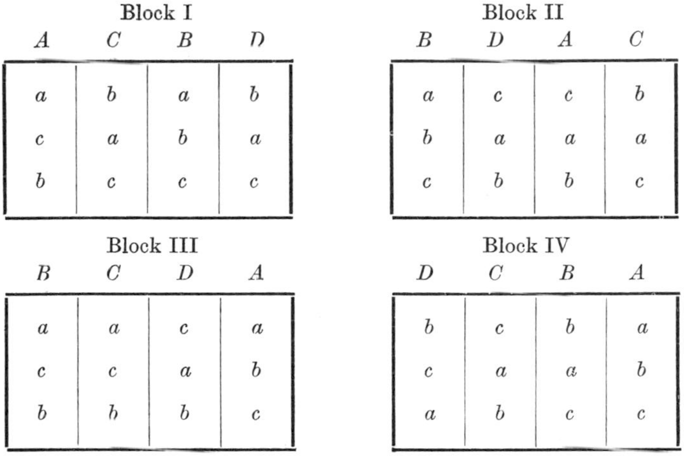

Block I

A	*C*	*B*	*D*
a	*b*	*a*	*b*
c	*a*	*b*	*a*
b	*c*	*c*	*c*

Block II

B	*D*	*A*	*C*
a	*c*	*c*	*b*
b	*a*	*a*	*a*
c	*b*	*b*	*c*

Block III

B	*C*	*D*	*A*
a	*a*	*c*	*a*
c	*c*	*a*	*b*
b	*b*	*b*	*c*

Block IV

D	*C*	*B*	*A*
b	*c*	*b*	*a*
c	*a*	*a*	*b*
a	*b*	*c*	*c*

Fig. 1. Main treatments: *A*, *B*, *C*, *D*. Sub-treatments: *a*, *b*, *c*.

plots and sub-treatments, may be divided into two parts, namely that due to interaction between sub-treatments and main plots having the same main treatment and that due to interaction between sub-treatments and main plots having different treatments. The second variance forms an estimate of the interaction of sub-treatments with main treatments, and this may be compared with the first variance, which forms a valid estimate of error also for the comparison of average sub-treatment differences. The second part of the analysis of variance is therefore as follows:

	Degrees of freedom
Main plots (= total of 1st part) ...	15
Sub-treatments...	2
Interaction: sub- and main treatments	6
Sub-plot error	24
Total	47

Experiments of this nature, although they are practically convenient, are not very efficient. Sub-plot error, representing chance differences between closely contiguous plots, is in general likely to be considerably smaller than the main-plot error, which is subject to all the causes of variation which affect the sub-plots (except such as arise from certain competition effects, unequal division of plots, and the like), and also to additional causes which affect the individual main plots as a whole. The accuracy of the comparison of sub-treatments, and of the interaction of main treatments with sub-treatments, is increased over that which would be obtained from a simple four-block experiment with 12 plots per block, but only at the expense of the comparison between the main treatments. In certain circumstances this may be what is required, but in general the direct effects are of more interest than the interactions.

This type of confounding is capable of many variations. The split plot experiment, where one half of each plot receives a different treatment from the other half, this difference being superimposed on the main treatments, is an example. In general the design and analysis of such experiments is comparatively simple, and if the sub-treatments are properly randomised within each separate main plot the conclusions are of certain validity.

A variant on the same theme is illustrated in Fig. 2. Here the treatments *a*, *b*, *c*, are not arranged at random within each main plot, but are restricted so as to lie in strips across the whole experiment. In the figure shown the main treatments *A*, *B*, *C*, *D*, are arranged in the form of a Latin square, though the same type of arrangement would be equally applicable to a randomised block experiment. Here the analysis of variance must be divided into three parts, with errors appropriate to comparisons of main treatments, to strip treatments, and to interactions respectively. The first part consists of an ordinary Latin square analysis, the second a randomised block analysis, each column of main plots forming one block of strips, the third a sub-plot analysis with the effect of strips as well as main plots removed. The analysis can be set out as in Table VII.

Table VII.

Main treatments	D.F.	Strip treatments	D.F.		Interactions	D.F.
Rows	3	Columns ...	3		Main plots	15
Columns ...	3	Strip treatments	2	} = {	Deviations of strips from columns ...	8
Main treatments	3	Error	6		Interactions ...	6
Error	6				Error	18
Main plots ...	15	Strips	11		Total	47

This type of lay-out is given here because it has certain practical attractions. It is open to even stronger criticism on the score of efficiency than the foregoing type, since neither main nor strip treatments are as accurately determined as are the interactions.

a	*c*	*b*	*c*	*a*	*b*	*b*	*a*	*c*	*c*	*b*	*a*
	A			*C*			*B*			*D*	
	B			*D*			*A*			*C*	
	C			*B*			*D*			*A*	
	D			*A*			*C*			*B*	

Fig. 2.

There is one type of design which at first appears to be a simple extension of the principle already employed, but for which the ordinary analysis breaks down. If a Latin square has whole rows (or columns) subjected to different treatments, trouble arises from the fact that the interactions of the Latin square (or sub-) treatments with the main row treatments are not orthogonal with the column effects. Fig. 3 shows a lay-out of this description.

A	*b*	*d*	*e*	*a*	*c*	*f*
B	*e*	*a*	*b*	*f*	*d*	*c*
B	*a*	*b*	*d*	*c*	*f*	*e*
A	*c*	*e*	*f*	*d*	*a*	*b*
B	*d*	*f*	*c*	*b*	*e*	*a*
A	*f*	*c*	*a*	*e*	*b*	*d*

Fig. 3.

The effects of the main treatments, *A* and *B*, are clearly obtained by an analysis in the form of three randomised blocks, 1 degree of freedom being allotted to treatments, 2 to blocks, and 2 to error. The experiment is of course not capable of giving a very precise answer on this difference. The analysis of the sub-treatments, on similar lines to that previously adopted, is given in Table VIII.

Table VIII.

	Degrees of freedom
Rows (= main plot total)	5
Columns	5
Sub-treatments	5
Interaction: main and sub-treatments	5
Sub-plot error	15
Total	35

When we come to the computation of the interaction sum of squares a difficulty presents itself. This interaction would ordinarily be computed as the interaction of Table IX.

Table IX.

Main treatments	Sub-treatments					
	a	*b*	*c*	*d*	*e*	*f*
A	—	—	—	—	—	—
B	—	—	—	—	—	—
Sum ($A+B$)	—	—	—	—	—	—
Difference ($A-B$)	—	—	—	—	—	—

The interaction involves a comparison of the differences ($A-B$). But on consideration it will be seen that these differences are not equalised for columns, the difference for *a*, for example, being made up of the sum of plots from columns 3, 4 and 5, less plots from columns 1, 2 and 6, that for *b* by plots from columns 1, 5 and 6, less plots from columns 2, 3 and 4. The comparison will therefore be affected not only by interaction differences, but also by column differences, and if the interactions are themselves negligible the sum of squares will tend to be increased. If therefore, as is usual, the error sum of squares is computed as a difference from the total sum of squares, the error will tend to be too small.

It will be seen that the sub-treatment sum of squares may be computed from the sum line ($A+B$) of the table, since this is equalised for both rows and columns. The sub-treatments are therefore orthogonal with the rows and columns, this being the ordinary Latin square property. The interactions are orthogonal to the rows, since each difference ($A-B$) is made up of the sum of plots from rows 1, 4 and 6, less plots from rows 2, 3 and 5.

To make a full analysis of such an experiment it is necessary to evaluate the interaction sum of squares, and this can only be done by fitting suitable constants to represent these interactions, together with constants for the columns, with which the interactions are not orthogonal. The total sum of squares attributable to the 10 degrees of freedom, columns and interactions, is thus obtained, and the error sum of squares is found by deducting this and the row and treatment sums of squares from the total sum of squares. The interaction sum of squares is given by the difference of the sum of squares for the 10 degrees of freedom above and the sum of squares for columns. The whole process is analogous to that already considered in section III.

Since this process necessitates the solution of ten simultaneous linear equations (or, on reduction, five equations) the amount of computation is vastly greater than what would be required in an orthogonal experiment of similar size. This type of lay-out should therefore be avoided. It is worth noting, however, that an approximate analysis can be made on the assumption that the interactions are negligible, the 20 degrees of freedom for interaction and sub-plot error being combined and utilised as an estimate of error for sub-treatments. This procedure is justified by the fact that in agricultural experiments interactions are generally found to be small in comparison with main effects. Should the assumption not be true the computed error will be too large and the experiment will be judged to be less efficient than it really is, but no false conclusions will be reached.

The Rothamsted experiment on sugar beet already mentioned was a somewhat complex example of a lay out of this type, columns as well as rows being subjected to additional treatments. This criss-cross pattern introduces a further complication which will be discussed in detail in the next section, which contains an outline of the correct analysis of the experiment.

V. The Rothamsted experiment on sugar beet in 1929.

A general description of this experiment has been published in *Arch. für Pflanzenbau*(1), but since this paper may not be readily accessible it will be as well to recapitulate the essential features here.

The experiment was originally designed as a 12×12 Latin square, the twelve treatments being made up of three nitrogen treatments, two of potash, and two of common salt, in all combinations. In addition it was desired to investigate the effect of phosphate, and to carry out the experiment on two varieties instead of one. These new factors were introduced into the original design by the artifice of sowing the rows of the Latin

	P	O	O	P	O	P	O	P	P	O	O	P	Row totals
J	9 15·1	3 1·0	1 – 7·6	8 2·6	4 –15·5	6 1·9	12 – 2·3	2 – 3·2	10 – 7·3	11 3·7	7 6·3	5 29·8	24·5
K	10 6·0	5 9·3	9 7·5	2 – 8·0	3 – 7·1	4 –10·2	8 – 4·5	1 –17·8	11 – 4·5	7 –17·8	12 0·2	6 2·9	– 44·0
J	11 28·6	1 4·6	7 8·9	12 20·6	10 6·2	3 23·7	5 7·9	4 6·8	8 8·1	6 4·4	9 8·6	2 10·6	139·0
K	6 21·1	12 9·8	2 17·1	3 17·9	9 13·7	7 9·5	10 7·8	8 13·0	1 – 1·9	4 9·2	5 12·2	11 14·9	144·3
K	8 23·5	11 19·7	6 19·7	4 0·6	1 – 3·5	5 8·3	2 2·8	9 1·6	7 0·9	10 4·5	3 – 2·1	12 5·0	81·0
J	3 28·4	7 2·8	10 – 5·1	6 10·1	5 8·7	9 15·6	11 3·3	12 7·7	4 – 3·0	2 9·8	1 –12·3	8 7·8	73·8
K	2 7·3	9 13·6	4 3·3	10 – 4·7	7 3·3	12 18·5	1 – 3·5	11 7·0	6 2·7	5 – 5·6	8 – 0·2	3 – 3·4	38·3
J	7 11·8	6 13·6	12 18·1	1 11·8	8 14·1	2 23·8	9 5·7	10 – 4·1	5 – 0·7	3 1·0	11 – 0·8	4 – 7·9	86·4
J	5 17·1	2 10·4	3 14·7	7 14·0	12 15·5	11 29·6	4 1·1	6 10·6	9 8·8	8 8·1	10 – 2·3	1 – 1·2	126·4
K	1 – 1·2	4 –10·5	8 1·7	5 – 0·4	11 – 5·8	10 – 9·8	7 5·8	3 – 2·4	2 5·0	12 4·8	6 9·3	9 1·2	– 2·3
J	12 16·2	8 10·7	5 3·5	11 12·3	2 – 6·3	1 5·9	6 16·3	7 10·6	3 7·8	9 13·1	4 7·7	10 10·9	108·7
K	4 –27·2	10 –27·0	11 –24·4	9 –24·4	6 –28·4	8 –21·0	3 –20·7	5 –12·2	12 – 5·8	1 –18·9	2 –15·3	7 –12·2	–237·5
Column totals	146·7	58·0	57·4	52·4	–5·1	95·8	19·7	17·6	10·1	16·3	11·3	58·4	538·6

Key to treatments and varieties.

Treatment...	1	2	3	4	5	6	7	8	9	10	11	12
Sulphate of ammonia	–	×	–	–	×	–	–	×	–	–	×	–
Nitrate of soda	–	–	×	–	–	×	–	–	×	–	–	×
Muriate of potash	–	–	–	×	×	×	–	–	–	×	×	×
Salt	–	–	–	–	–	–	×	×	×	×	×	×

O, No superphosphate. *P*, superphosphate. *J*, Kuhn (Johnson's Perfection). *K*, Kleinwanzleben.

square with the two varieties, one of each pair of rows being allotted at random to the first variety; and similarly by treating one of each pair of columns with phosphate.

The details of the lay-out and the yields of roots per plot are given in Fig. 4. The analysis of variance, assuming the interactions of phosphate and variety with the other manurial treatments (33 degrees of freedom) are negligible, and can be classed as error, is given in Table X. All the sets of degrees of freedom classified in this table are orthogonal with one another and can be computed in the ordinary manner.

Table X. *Analysis of variance.*

		Degrees of freedom	Sums of squares	Mean square
Rows	Row pairs	5	4324·76	
	Variety	1	2328·06	2328·06
	Error	5	3651·27	730·25
		11	10304·09	
Columns	Column pairs	5	865·94	
	Phosphate	1	346·58	346·58
	Error	5	500·70	100·12
		11	1713·22	
Treatments	Nitrogen	2	2008·47	1004·24
	Potash	1	5·60	5·60
	Salt	1	203·54	203·54
	Interactions: $K \times$ Na	1	34·22	34·22
	$N \times K$	2	192·47	96·24
	$N \times$ Na	2	97·47	48·74
	$N \times K \times$ Na	2	64·92	32·46
		11	2606·70	
Interaction: phosphate and varieties		1	234·60	234·60
Interactions: phosphate, variety, and other manures		33	4997·69	45·85
Error		76		
Total		143	19856·30	

The interactions between phosphate and other manurial treatments (11 degrees of freedom), between variety and other manurial treatments (11 degrees of freedom), and between phosphate, variety, and other manurial treatments (11 degrees of freedom) are all non-orthogonal with the rows and columns, and consequently the evaluation of the sum of squares to be allotted to them involves the fitting of appropriate constants. There will in all be 55 independent constants, and consequently 55 independent equations to solve. These equations can be immediately reduced by substitution to 22, which are fairly easily solved by iterative methods.

There is no need to give more than an outline of the analysis here. The constants were chosen according to the following scheme:

Mean	m;
Rows	$r_1, r_2, ..., r_{12}, (r_1 + r_2 + ... + r_{12} = 0)$;
Columns	$c_1, c_2, ..., c_{12}, (c_1 + c_2 + ... + c_{12} = 0)$;
Non-phosphatic manures ...	$t_1, t_2, ..., t_{12}, (t_1 + t_2 + ... + t_{12} = 0)$;
Interaction, phosphate and variety	g;
Interaction, phosphate and other manures	$p_1, p_2, ..., p_{12}, (p_1 + p_2 + ... + p_{12} = 0)$;
Interaction, variety and other manures	$v_1, v_2, ..., v_{12}, (v_1 + v_2 + ... + v_{12} = 0)$;
Interaction, variety, phosphate and other manures	$i_1, i_2, ..., i_{12}, (i_1 + i_2 + ... + i_{12} = 0)$.

The exact meaning assigned to the above constants will be clear from the following table (Table XI), showing the constants assigned to each of the 48 varietal and treatment combinations.

Table XI.

Treatments	1	2	etc.
JO	$t_1 - p_1 - v_1 + i_1 + g$	$t_2 - p_2 - v_2 + i_2 + g$	
JP	$t_1 + p_1 - v_1 - i_1 - g$	$t_2 + p_2 - v_2 - i_2 - g$	
KO	$t_1 - p_1 + v_1 - i_1 - g$	$t_2 - p_2 + v_2 - i_2 - g$	
KP	$t_1 + p_1 + v_1 + i_1 + g$	$t_2 + p_2 + v_2 + i_2 + g$	

Corresponding to the 144 plots there are 144 equations. If Y_{xyz} represents the yield of the xth row, the yth column, and the zth treatment, the equation corresponding to the fifth plot of the fourth row, for instance, is

$$Y_{459} = m + r_4 + c_5 + t_9 - p_9 + v_9 - i_9 - g.$$

The equations for determining the constants, deducible from these, are therefore

$$\begin{array}{llr}
144m = SY_{xyz}, & & 1 \text{ equation} \\
12m + 12r_a + S(\pm p)^* + S(\pm i) = SY_{ayz}, & & 12 \text{ equations} \\
12m + 12c_a + S(\pm v)^* + S(\pm i) = SY_{xaz}, & & 12 \quad ,, \\
12m + 12t_a = SY_{xya}, & & 12 \quad ,, \\
144g = S(\pm Y_{xyz}), & & 1 \quad ,, \\
12p_a + S(\pm r) + S(\pm c)^* = S(\pm Y_{xya}), & & 12 \quad ,, \\
12v_a + S(\pm r)^* + S(\pm c) = S(\pm Y_{xya}), & & 12 \quad ,, \\
12g + 12i_a + S(\pm r) + S(\pm c) = S(\pm Y_{xya}), & & 12 \quad ,,
\end{array}$$

The ± signs in these equations vary from equation to equation, and can only be determined by reference to the plan of the experiment. In the actual solution a key diagram of signs was prepared; the diagonal symmetry obviates the necessity of tabulating more than half the signs.

Only 68 of the 74 equations are independent. The equations for m, g and all t are free from entanglement and can be solved immediately. By means of the last 36 equations all p, v and i can be eliminated from the first two groups of 12 equations. The resultant 24 equations can then be solved by iteration. The solution converges rapidly. The fact that $S(r)$ and $S(c)$ are both zero gives two useful checks. It should be noted that the sets of terms marked with asterisks are the same for each equation of the set in which they occur, and can be evaluated numerically before substitution is begun.

After values had been obtained for all r, c, p, v and i, the reduction in the sum of squares due to fitting was calculated by means of the ordinary formula. As a check the error sum of squares was also computed directly from the residuals. The values obtained were 4015·38 and 4003·17 respectively. The agreement, though not perfect, is close enough to indicate that there is nothing seriously wrong with the fitting.

The last term of the previous analysis of variance splits up as in Table XII.

Table XII.

	Degrees of freedom	Sums of squares	Mean square
Rows, columns, and p, v, i interactions	55	13011·82	
Rows and columns	22	12017·31	
p, v, i interactions	33	994·51	30·1
Error	76	4003·17	52·7

The p, v, i interactions have a lower variance than the error variance, the difference lying outside the lower 5 per cent. point. This significantly lower variance is apparently a chance effect, but it may be that there is some property of the lay-out which decreases the expectation of variance in these interactions.

In any case there is little doubt that the high order interactions of this experiment are negligible. The original analysis, by neglecting the fact that the rows and columns were not orthogonal with the treatments, and working on the crude treatment totals, produced apparent large interaction sums of squares which were in reality due to row and column effects. This in turn reduced the error sum of squares much below its true value, giving an error variance of 29·09. As a consequence many of these interactions appeared to be significant, leading to conclusions of the type: "In the absence of nitrogen the response to phosphate only occurs if muriate of potash be absent also; in the presence of sulphate of ammonia the response to phosphate only occurs if either muriate of potash or salt is present; while in the presence of nitrate of soda the response to phosphate only occurs in the absence of salt" (*loc. cit.* pp. 581–2).

It will now be seen that the experiment, as it stands, is quite capable of giving a clear and unambiguous verdict on the main manurial effects, and that to obtain this verdict it is only necessary to carry out the first part of the analysis, combining the 33 non-orthogonal interaction degrees of freedom with the 76 error degrees of freedom. The extraction of the 33 interaction degrees of freedom has only confirmed the approximate procedure. But the experiment is not suitable for providing any estimate of the interactions between variety or phosphate and the non-phosphatic manures. An intolerable amount of labour would be required if it were desired to split up the 33 degrees of freedom further and examine individual interactions of this series, and although in this particular case the very low value of the sum of squares for the 33 degrees of freedom and a general examination of the values of the interaction constants (not given here) indicates that there is no significant interaction of this kind, yet in other cases it might well be that one or two of the principal interactions of the series might exist and be large enough to be significant. But the existence of a few such interactions, unless they were very large, would not appreciably affect the approximate error, owing to the number of degrees of freedom involved. On the other hand any definite verdict of significance or non-significance of these interactions themselves would be very troublesome.

The possibility of designing an experiment, similar to this one, but avoiding its disadvantages, must now be considered. Some sacrifice of the number of manurial treatments appears to be necessary. In the arrangement illustrated in Fig. 5 the manurial treatments 1–6 are arranged in the form of a 6 × 6 Latin square. Each row of this Latin square is subdivided into two strips receiving treatments A_1 and A_2, and similarly each column is divided into two strips receiving treatments B_1 and B_2. A_1 and A_2, and B_1 and B_2, are assigned at random within each pair of strips.

The analysis can be set out as in Table XIII. There are six separate and distinct errors appropriate to comparisons involving row strips, column strips, the Latin square plots, differences between main-plot halves (two types), and cross differences between the main-plot quarters.

This analysis is of particular interest in that it shows up a defect, which has not yet been discussed, in the analysis of the sugar-beet experiment by the method of fitting constants. It has been tacitly assumed that the 76 degrees of freedom for error in that experiment are homogeneous, and that any set of treatment degrees of freedom, except the direct phosphate and varietal effects, may be validly compared with them. For the new lay-out there are no less than 4 error sets of degrees of freedom corresponding to these 76 degrees of freedom and to the 24 additional error degrees of freedom introduced by reducing the number of treatments from 12 to 6. Although perhaps no great difference between the last three errors is to be expected, since they all depend on differences within the large plots, there is no question that the Latin square error will tend to be larger than the other three. Any analysis which confuses these errors will be of doubtful validity.

Table XIII.

Row strips	D.F.	Column strips	D.F.	Latin square	D.F.
Rows	5	Columns	5	Rows	5
$A_1 \, v. \, A_2$	1	$B_1 \, v. \, B_2$	1	Columns	5
Error	5	Error	5	Treatments	5
	—		—	Error	20
Row strips	11	Column strips	11		—
				Total	35
Differences between large plot halves: $A_1 - A_2$		Differences between large plot halves: $B_1 - B_2$		Differences between large plot quarters: $A_1B_1 - A_1B_2 - A_2B_1 - A_2B_2$	
	D.F.		D.F.		D.F.
Deviations of row strips from rows	6	Deviations of column strips from columns	6	Interaction $A \times B$	1
Interactions $A \times 1-6$	5	Interactions $B \times 1-6$	5	Interactions $A \times B \times 1-6$	5
Error	25	Error	25	Error	30
	—		—		—
Total	36	Total	36	Total	36

This point throws light on the question of the direct interaction of variety and phosphate. It seems probable that a specially accurate comparison should be available for this particular interaction, since it involves only differences of the type $PK - PJ - OK + OJ$, and plots receiving these four treatments occur in sets of four throughout the experiment. In the original analysis by Dr Wishart a special estimate of error was made for this interaction by comparing the variance of the 36 tetrad differences with the mean value of these differences, after compensating for effects of treatments 1–12 by applying adjustments to the individual plot yields, equal in magnitude to the corresponding treatment means. The values obtained were as follows.

	Degrees of freedom	Sums of squares	Mean squares
Interaction of phosphate and varieties	1	234·60	234·60
Deviations of tetrad differences	35	547·30	15·64

This estimate of error variance is much below the general estimate for the 76 degrees of freedom, namely 52·7, and though the method of compensating for treatment effects introduces certain correlations which will tend to lower somewhat the estimate from tetrad differences, there is no doubt that the two errors are in fact significantly different. This difference offers a striking practical confirmation of the non-homogeneity of the 76 degrees of freedom for error used in the main analysis.

	B_2	B_1	B_1	B_2	B_2	B_1	B_2	B_1	B_1	B_2	B_2	B_1
A_1	1	1	3	3	2	2	6	6	4	4	5	5
A_2	1	1	3	3	2	2	6	6	4	4	5	5
A_1	5	5	4	4	1	1	3	3	2	2	6	6
A_2	5	5	4	4	1	1	3	3	2	2	6	6
A_2	3	3	5	5	6	6	2	2	1	1	4	4
A_1	3	3	5	5	6	6	2	2	1	1	4	4
A_1	6	6	1	1	5	5	4	4	3	3	2	2
A_2	6	6	1	1	5	5	4	4	3	3	2	2
A_2	4	4	2	2	3	3	5	5	6	6	1	1
A_1	4	4	2	2	3	3	5	5	6	6	1	1
A_2	2	2	6	6	4	4	1	1	5	5	3	3
A_1	2	2	6	6	4	4	1	1	5	5	3	3

Fig. 5.

The lay-out given in Fig. 5 suffers from the disadvantage that the comparisons of the treatments 1–6 involve the large plots of the 6×6 Latin square, so that the maximum efficiency is not obtained. By confounding interactions between A, B and 1–6 with the large plots of the Latin square, in place of the treatments 1–6, a gain in efficiency on the direct comparisons may be obtained, but this confounding is only possible with six sub-treatments when there are six sub-plots within each Latin

	B_2	B_1	B_2	B_1	B_1	B_2	B_2	B_1
A_1	2	1	1	2	3	4	3	4
	α		β		γ		δ	
A_2	4	3	3	4	1	2	1	2
A_2	1	2	4	3	4	3	2	1
	δ		α		β		γ	
A_1	3	4	2	1	2	1	4	3
A_2	3	4	2	1	2	1	4	3
	β		γ		δ		α	
A_1	1	2	4	3	4	3	2	1
A_1	4	3	3	4	1	2	1	2
	γ		δ		α		β	
A_2	2	1	1	2	3	4	3	4

Fig. 6.

square plot, necessitating three A (or B) treatments and 216 plots. Fig. 6 illustrates a lay-out of this type, but with four sub-treatments only, giving in all 64 plots. Each large plot of the Latin square consists of four plots receiving treatments 1–4. The conjunction of the treatments 1–4 and the A and B treatments within the plots of the Latin square is determined by a subsidiary Latin square:

Main plot	A_1B_1	A_1B_2	A_2B_1	A_2B_2
α	1	2	3	4
β	2	1	4	3
γ	3	4	1	2
δ	4	3	2	1

This has the effect of confounding the interaction α *v.* β *v.* γ *v.* δ with the Latin square plots, while keeping the strips orthogonal with treatments 1–4. (This confounding of high-order interactions is dealt with more fully in the next section). If treatments 1, 2, 3 and 4 represent the four combinations P_1Q_1, P_2Q_2, P_2Q_1, P_1Q_2 respectively of two pairs of treatments P_1, P_2, and Q_1, Q_2, the analysis of Table XIV is obtained, the interactions $\alpha\, v.\, \beta\, v.\, \gamma\, v.\, \delta$ being equivalent to $B\times Q$, $A\times P\times Q$ and $A\times B\times P$, etc.

Table XIV.

Row strips	D.F.	Column strips	D.F.	Latin square	D.F.
Rows	3	Columns	3	Rows	3
A_1 v. A_2	1	B_1 v. B_2	1	Columns	3
Error	3	Error	3	$B\times Q$	1
	—		—	$A\times P\times Q$	1
Row strips	7	Column strips	7	$A\times B\times P$	1
				Error	6
				Total	15
Plot halves, A_1-A_2		Plot halves, B_1-B_2		Plot quarters	
Deviations of row		Deviations of column		$A\times B$	1
strips from rows	4	strips from columns	4	P	1
$P\times Q$	1	Q	1	$A\times Q$	1
$B\times P$	1	$A\times P$	1	$B\times P\times Q$	1
$A\times B\times Q$	1	$A\times B\times P\times Q$	1	Error	12
Error	9	Error	9		—
	—		—	Total	16
Total	16	Total	16		

It should be noticed that the subsidiary Latin square given is the only one for which the interactions $\alpha\, v.\, \beta\, v.\, \gamma\, v.\, \delta$ represent real physical effects. An alternative is to take 1–4 to represent P_1Q_1, P_1Q_2, P_2Q_1 and P_2Q_2 respectively, when $\alpha\, v.\, \beta\, v.\, \gamma\, v.\, \delta$ will be equivalent to $A\times P$, $B\times Q$ and $A\times B\times P\times Q$. Larger experiments can be designed on similar lines. The 6×6, 8×8 and 9×9 squares have been investigated, and present no theoretical difficulty.

VI. The confounding of interactions.

A type of confounding radically different to that considered in the last two sections occurs when the experiment is so arranged that degrees of freedom corresponding to high-order interactions are confounded with block differences. This type of confounding does not contribute directly to ease of execution, but it very greatly increases the accuracy of complex experiments.

It will be recalled(2) that a great gain in efficiency is attained by carrying out the examination of several effects in a single experiment. If, for example, it is desired to examine the manurial response of a crop to nitrogen, phosphate and potash, a single experiment made up of plots receiving all possible combinations of the three manures will be much more efficient, other things being equal, than three separate experiments each of one-third the size. If, however, several rates of application for each manure are included the number of plots required for each complete replication becomes very large, and the size of blocks in a randomised block experiment must be correspondingly increased; the magnitude of soil differences of plots within the same block may then become unduly great, with the result that soil heterogeneity is not successfully eliminated by the blocks.

This difficulty can be overcome by sacrificing complete replication within each block. If, for instance, the plots of each completely replicated block are divided into a set of sub-blocks in such a manner that the comparison between sub-block totals corresponds to the comparison representing a set of high-order interactions, then the degrees of freedom corresponding to this set of interactions will be confounded with sub-block differences. Moreover each completely replicated block may be divided so that different high-order interactions are confounded, and thus some information is obtained on every degree of freedom, those that are confounded in one set of blocks being kept clear of block differences in other sets. In such a case it may well happen, if the accuracy of the comparisons not confounded is considerably increased, that even those that suffer some measure of confounding are more accurately determined than would be the case in a straightforward experiment.

As an example of complete confounding of a single degree of freedom corresponding to a high order interaction the type of experiment in which two levels of three different treatments, such as nitrogenous, phosphatic, and potassic fertilisers, are arranged in all possible combinations, eight in number, may be considered. There are seven degrees of freedom for treatments, which can be divided up into three for direct effects, three for first order interactions, and one for the second order interaction. The sum of squares corresponding to each single degree of freedom may be computed from the sum of half the plots less the sum of the other half; the direct nitrogen effect is measured by the difference of the sums of all the plots receiving nitrogen and those receiving no nitrogen, the interaction of nitrogen and potash from the difference of the sums of all the plots receiving both nitrogen and potash, or neither, and those receiving one or the other, the second order interaction from the difference of the sum of all the plots receiving all three manures, or one only, and the sum of those receiving two manures, or none. Bearing this in mind, it will be seen that the division of each complete replication into two blocks as in Table XV, types I*A* and I*B*, will leave every comparison free of block differences except the second order interaction, which will be completely confounded with block differences. The full analysis of variance in a 32-plot experiment is given in Table XVI.

Table XV. *Block types in* $2 \times 2 \times 2$ *experiment.*

I*A*	I*B*	II*A*	II*B*	III*A*	III*B*	IV*A*	IV*B*
N	*O*	*N*	*O*	*N*	*O*	*P*	*O*
P	*NP*	*P*	*K*	*K*	*P*	*K*	*N*
K	*NK*	*NK*	*NP*	*NP*	*NK*	*NP*	*PK*
NPK	*PK*	*PK*	*NPK*	*PK*	*NPK*	*NK*	*NPK*

Table XVI.

	D.F.
Second order interaction $N \times P \times K$	1
Block pairs	3
Error	3
Blocks	7

		D.F.
Direct effects	*N*	1
	P	1
	K	1
First order interactions	$N \times P$	1
	$N \times K$	1
	$P \times K$	1
Blocks		7
Error		18
Total		31

The first part of the analysis assumes that the two types of block are actually grouped in pairs. In practice this part is not likely to be of any value; it is set out here for the sake of logical completeness only.

In this design all the information on the second order interaction (except the negligible amount derivable from the comparison of block totals) is sacrificed. The complete set of types in Table XV gives an alternative arrangement in which one-quarter of the information on each of the interaction degrees of freedom is sacrificed. The pairs of blocks I*A* and I*B*, II*A* and II*B*, III*A* and III*B*, IV*A* and IV*B*, each contain a complete replication, the pairs II, III and IV being split so as to confound the first order interactions $N \times P$, $N \times K$ and $P \times K$, respectively. The partition of the degrees of freedom will now be:

Direct effects	3
First order interactions	3
Second order interaction	1
Blocks	7
Error	17
Total	31

In this experiment the computation of the sums of squares to be allotted to the degrees of freedom which are partially confounded presents no difficulty. The second order interaction, for example, is simply computed from the blocks II*A*, II*B*, III*A*, III*B*, IV*A*, IV*B*, with which it is orthogonal, omitting entirely blocks I*A* and I*B*. The sum of squares allotted to error can, as usual, be found by subtraction from the total sum of squares. It should be particularly noticed that this method of computation depends on the fact that the degree of freedom in question is *completely* confounded with the omitted pair of blocks, and orthogonal with the remaining blocks.

Yet another opportunity of obtaining information on degrees of freedom that at first sight appear to be confounded with block differences occurs when each sub-block contains a plot or plots receiving the same treatment as corresponding plots in the other sub-blocks, for then the differences between these plots furnish a measure of sub-block differences. If, for example, one block contained plots receiving treatments A and B, and another block contained plots receiving treatments A and C, it would be possible to make a comparison of B and C by comparing B with A in the first block and C with A in the second block.

As an example of this type of confounding, take the system of treatments, nitrogen and no nitrogen, applied early or late, and phosphate and no phosphate, eight combinations in all, of which the pair phosphate without nitrogen applied early or late, and the corresponding no phosphate pair, are identical. This system corresponds exactly to the system already considered, early and late application (of nitrogen) being substituted for potash. The experiment is not very efficient on the question of time of application, since only half the plots enter into this comparison, but this is unavoidable in experiments of this type: some increase of efficiency might be obtained by retaining only one of each pair of identical plots, but at the expense of balance on the phosphate and nitrogen comparisons.

If the plots of each replication are divided into two sub-blocks according to type I of the previous example we shall have block pairs of the type:

O_1	O_2
P_1	P_2
$N\ (L)$	$N\ (E)$
$PN\ (E)$	$PN\ (L)$

The five degrees of freedom for treatments may be partitioned as follows:

Phosphate	1
Nitrogen	1
Time of application	1
$P \times N$	1
$P \times T$	1

At first $P \times T$, which would ordinarily be computed from the sum of quantities of the type $N\ (E) + PN\ (L) - N\ (L) - PN\ (E)$, appears to be confounded with blocks, but, as explained above, a comparison can be made by means of the other plots. The sum of squares is in fact (for complete fourfold replication)

$$\tfrac{1}{32}\left[S\left\{N(E) + PN(L) - N(L) - PN(E) - O_2 - P_2 + O_1 + P_1\right\}\right]^2.$$

This is orthogonal with the blocks and with all the other treatment degrees of freedom.

The above examples have purposely been chosen for their simplicity. The full enumeration of the different types of confounding in even the simpler types of lay-out is necessarily very complex and has never been fully investigated; but the examples given appear to embody the principles involved. Attention has been confined to cases where each full set of replicated plots is divided into two parts only, thus confounding a single degree of freedom. Extension to the case where there is division into more than two parts, with the consequent confounding of a set of degrees of freedom, presents no particular difficulty beyond that of choosing a suitable type of division.

The possibilities of confining the confounding to high-order interactions appear to be very limited. In an experiment consisting of three levels of three different treatments in all combinations, making 27 treatments in all, it is possible to confound two of the eight-second order interactions by splitting into three blocks of nine plots, and by varying the manner of splitting the whole 8 degrees of freedom may be partially confounded (as in the *NPK* example), but it is not possible to confound the whole 8 degrees of freedom by splitting into nine blocks of three plots. The best that can be done in this way is to confound degrees of freedom corresponding to two-second order and six first-order interactions.

An example of neat adaptation of the principle of confounding to a special purpose is afforded by the series of experiments on potatoes at Rothamsted begun in 1925 and ended in 1931. The first three experiments of this series have already been discussed in a previous paper in this *Journal*(3). The principle of confounding was first resorted to in the 1927 experiment, though at that time the possibility of gaining useful information on confounded degrees of freedom was not realised, and the 1927 experiment has a certain lack of symmetry which greatly complicates the full analysis. An outline of the numerical computation of the 1931 experiment is given here, as providing an excellent illustration of the simplicity of the analysis of apparently complex experiments when their design is satisfactory.

It will be recalled that the purpose of these experiments was to investigate the effect of nitrogenous and potassic fertilisers on potatoes.

In addition to the ordinary quantitative effects it was desired to compare three different sources of potash, namely sulphate of potash, muriate of potash, and low grade potash manure salt. The lay-out of the experiment, which is very similar to the 1927 experiment, is given in Fig. 7, together with the yields of the individual plots. In addition to the treatments shown the plots were split, one half receiving a dressing of superphosphate. There is no need to repeat this part of the analysis here, as it is quite straightforward, consisting in essence of an analysis of the differences of pairs of half-plots; as in this type of analysis no account is taken of blocks, it is unaffected by the confounding.

The full analysis of variance is given in Table XVII. The total sum of squares, that for blocks, and that for all treatment effects except the interactions involving both nitrogen and potash quality, are computed in the ordinary manner, except that the quality effect has been computed from the weighted means of the various qualities, giving double weight to the plots receiving a double dressing of potash. The sum of squares is in fact the sum of the squares of the deviations of the totals

$$S\,(S_1 + 2S_2),\quad S\,(M_1 + 2M_2),\quad S\,(P_1 + 2P_2),$$

from their mean, divided by 45, *i.e.* $9 \times (1^2 + 2^2)$. The interaction orthogonal to this pair of degrees of freedom is that given by the differences of

$$S\,(2S_1 - S_2),\quad S\,(2M_1 - M_2),\quad S\,(2P_1 - P_2),$$

the sum of the squares of the deviations being, as before, divided by 45. It is important to notice that a change in the method of measurement of a direct effect necessarily involves a corresponding change in the interaction.

Table XVII. *Analysis of variance (whole plots)*

	Degrees of freedom	Sum of squares	Mean square
Blocks	8	5,006,460	
Nitrogen	2	1,215,036	607,518
Potash quantity	2	41,164	20,582
Potash quality	2	69,496	34,748
Nitrogen × potash quantity	4	62,528	15,632
Potash quantity × quality	2	125,176	62,588
Nitrogen × potash quality Nitrogen × quantity × quality	8	227,188	28,398
Error	52	1,641,848	31,574
Total	80	8,388,896	
Interactions: (*a*) for K_1	2	86,064	
(*b*) for K_1	2	40,988	
(*a*) for K_2	2	57,600	
(*b*) for K_2	2	42,536	
	8	227,188	

Author's note — Corrected table; in the original the sums of squares and mean squares were in error by a factor of 4

The three complete replications of the experiment are all divided differently into sub-blocks, and it would therefore at first sight appear that three pairs of degrees of freedom were confounded with blocks.

	A			B			C		
	3 1996	4 *P* 1638	5 *M* 1859	9 *M* 1978	2 2088	5 *P* 2197	1 2016	4 *M* 2398	6 *P* 2330	
	8 *P* 1818	7 *M* 1806	2 1972	1 1978	4 *S* 2185	6 *M* 2499	8 *M* 2535	3 2838	2 2382	
	6 *S* 2076	9 *S* 2115	1 2003	8 *S* 2460	3 2591	7 *P* 2655	9 *P* 2792	7 *S* 2649	5 *S* 2562	
	8 *S* 1687	7 *P* 1681	5 *S* 1822	5 *M* 2062	8 *M* 2304	2 2327	9 *S* 2384	3 2581	1 2122	
D	6 *P* 1623	9 *M* 1841	4 *M* 1586	6 *S* 2012	4 *P* 1942	9 *P* 2368	5 *P* 2203	8 *P* 2345	4 *S* 1996	F
	2 1651	1 1634	3 1665	7 *S* 1854	1 2064	3 2388	2 2195	6 *M* 2391	7 *M* 1685	
	4 *S* 1511	5 *P* 1687	9 *P* 1882	7 *M* 1877	9 *S* 2379	3 2462	4 *P* 2081	2 2288	6 *S* 2077	
	7 *S* 1562	3 1776	6 *M* 2080	4 *M* 1982	8 *P* 2082	6 *P* 2280	9 *M* 2296	8 *S* 2308	7 *P* 1652	
	8 *M* 1595	1 1501	2 1860	1 2096	5 *S* 2051	2 1889	3 2165	5 *M* 2007	1 1625	
	G			H			I			

Fig. 7.

Key to treatments.

Treatment	1	2	3	4	5	6	7	8	9
Nitrogen	0	1	2	0	1	2	0	1	2
Potash	0	0	0	1	1	1	2	2	2

S = sulphate of potash, M = muriate of potash, P = potash manure salt.

Actually, however, any one of these pairs can be constructed from the other two pairs, so that only 4 separate degrees of freedom are confounded.

The method of computation of the confounded degrees of freedom utilises the principle that the interaction of a 3 × 3 table (4 degrees of freedom) can be split up into two pairs of degrees of freedom. The confounded degrees of freedom are those involving the interaction of nitrogen with potash quality and with potash quality and quantity. These are equivalent to the interaction of nitrogen with potash quality for the single dressing of potash, and the similar interaction for the double dressing. The 4 degrees of freedom for the single dressing can be split up into two sets involving comparisons of

(*a*) $S_1N_0 + M_1N_1 + P_1N_2,\ M_1N_0 + P_1N_1 + S_1N_2,\ P_1N_0 + S_1N_1 + M_1N_2,$

and

(*b*) $S_1N_0 + P_1N_1 + M_1N_2,\ P_1N_0 + M_1N_1 + S_1N_2,\ M_1N_0 + S_1N_1 + P_1N_2.$

It will be seen that the first pair of comparisons is orthogonal with blocks, and consequently the sum of squares may be computed in the ordinary manner. The second pair consists of the comparison of all the single potash plots in blocks B, F and G with those in A, E and I and in C, D and H. To eliminate block effects the other plots in the block are utilised as already explained, the comparison being made between quantities of the type

$$S\,(2K_1 - K_2 - K_0),$$

the summations being taken over the three sets of blocks. The numerical values are

Blocks	$S\,(2K_1 - K_2 - K_0)$
B, F, G	+ 260
A, E, I	− 1841
C, D, H	− 888

The sum of the squares of the deviations of these numbers from their mean, divided by 54, *i.e.* $9 \times (2^2 + 1^2 + 1^2)$, gives the required sum of squares. For the double dressing of potash the computation is similar, but here the opposite pair of interactions is confounded.

It is important to notice that the validity of the above process depends on the fact that after compensation is made for block differences each pair of confounded degrees of freedom is not only orthogonal with blocks, but also with all treatment degrees of freedom, including the other confounded pair. A rigorous proof of the whole procedure can be made by fitting constants.

This orthogonal property depends on the symmetry of the experiment. It does not hold for the 1927 experiment, and consequently the above method of computation breaks down. A full analysis can best be made by fitting constants for the blocks and confounded interactions, and solving the resultant equations, a somewhat lengthy process; alternatively the confounded degrees of freedom can be combined with the error degrees of freedom, on the assumption that the effects if any are small, and consequently unlikely seriously to increase the estimate of error above its true value. This latter process was the one actually adopted in the original analysis, and is justified *a posteriori* in every other experiment of the series, but it is interesting to note that full analysis of the 1927 experiment revealed a very strong interaction between nitrogen and potash quality, this being due to the low yields of potash manure salt in conjunction with nitrogen. This effect can easily be seen if the treatment effects are judged from totals of deviations from block means instead of from crude totals; it accounts for the definitely significant differences between the different kinds of potash in this year. The effect is entirely absent in other years.

One further point should be mentioned in connection with the computation of confounded experiments. The differences between the crude treatment means corresponding to confounded degrees of freedom are affected by block differences. If the degrees of freedom are completely confounded no comparison is possible on the basis of the ordinary error; if the degrees of freedom are partially confounded comparison is still possible, but allowance must first be made for block differences, and it must be remembered that even then such comparisons are subject to higher error than the rest. Inasmuch as the analysis of variance provides precise tests of significance there is no need to evaluate special errors, but it is as well in exhibiting the results to avoid giving tables which involve completely confounded degrees of freedom. If it is desired to show an effect corresponding to a partially confounded degree of freedom the treatment means must be corrected. It must be remembered that the corrections cannot be deduced directly from the crude block means, since these means themselves contain differences due to the partially confounded degrees of freedom. It is therefore easiest to build up a table from the various treatment and interaction differences which have been determined by the analysis.

The 1931 potato experiment, of which the analysis has just been given, will serve as an example of the process. The estimates of the deviations from the mean yield produced by the nitrogen and potash in all combinations are one-ninth the deviations of the totals of these nine treatment combinations; the value for K_1N_0, for example, is $-148{\cdot}2$. The potash quality effects at the single level of potash are measured by one-ninth the deviations of the totals of the three qualities at this potash level from their mean, and a similar process serves for the double level of potash. The only remaining effects are the interactions of nitrogen with potash quality at the two levels of potash. At the single level of potash the (*a*) interactions are orthogonal with blocks, and the appropriate measures of the effects are therefore one-ninth the deviations of the totals of $S_1N_0 + M_1N_1 + P_1N_2$ and the two corresponding quantities, the first of these being $-$ 58·4. The (*b*) interactions are not orthogonal with blocks, and the appropriate measures are therefore one-eighteenth the deviations of the totals $S\ (2K_1 - K_2 - K_0)$ over the appropriate blocks. The first is $+$ 60·2. It will be noted that all these quantities can be immediately written down, since the appropriate totals have already been obtained in the analysis of variance. The estimation of a constant multiple of a single plot yield, in this case nine times, will further simplify the numerical work.

It remains to build up the final yield table. The entry for S_1N_0, for example, on a single plot basis, will be given by

$$2072{\cdot}6 - 148{\cdot}2 - 9{\cdot}7 - 58{\cdot}4 + 60{\cdot}2 = 1916{\cdot}5.$$

In conclusion it should be emphasised that it is by no means always necessary to construct tables of this type. In cases where the partially confounded degrees of freedom are not significant it usually suffices to exhibit the main unconfounded effects.

VIII. Summary.

The principle of orthogonality in replicated experiments is discussed and the dangers of non-orthogonality emphasised. The modifications necessary in the ordinary procedure of the analysis of variance when applied to non-orthogonal data are developed, attention being paid to the shorter methods, applicable in certain cases, by which the heavy labour of computation necessary in the general method of fitting constants may be avoided.

Certain modifications in the design of replicated experiments, usually designated by the term confounding, are explained. The different types of confounding are discussed, together with their uses, and the appropriate methods of analysis are set out. The methods are applied to the analysis of an experiment on sugar beet (where a previous incorrect analysis is corrected) and an experiment on potatoes.

In conclusion I desire to express my thanks to Dr R. A. Fisher for his helpful advice and criticism.

REFERENCES.

(1) WISHART, J. The analysis of variance illustrated in its application to a complex agricultural experiment on sugar beet. *Arch. für Pflanzenbau* (1931), **5**, 561.

(2) FISHER, R. A. and WISHART, J. The arrangement of field experiments and the statistical reduction of the results. *Imperial Bureau of Soil Science, Technical Communication* (1930), No. 10.

(3) EDEN, T. and FISHER, R. A. Studies in crop variation. VI. Experiments on the response to potash and nitrogen. *J. Agric. Sci.* (1929), **19**, 201.

(4) FISHER, R. A. The arrangement of field experiments. *J. Min. Agric.* (1926), **33**, 503.

(5) FISHER, R. A. *Statistical Methods for Research Workers* (1925). Edinburgh: Oliver and Boyd. (4th Edition, 1932.)

(6) ALLAN, F. E. and WISHART, J. A method of estimating the yield of a missing plot in field experimental work. *J. Agric. Sci.* (1930), **20**, 399.

(7) FISHER, R. A. Applications of Student's distribution. *Metron.* (1925), **5**, 1.

(8) IRWIN, J. O. Mathematical theorems involved in the analysis of variance. *J. Roy. Stat. Soc.* (1931), **94**, 284.

PAPER II

THE ANALYSIS OF REPLICATED EXPERIMENTS WHEN THE FIELD RESULTS ARE INCOMPLETE

FROM THE EMPIRE JOURNAL
OF EXPERIMENTAL AGRICULTURE
VOLUME I, No. 2, pp. 129–142, 1933

Author's Note

When some of the values are missing, or are discarded for some reason, the orthogonality of the original design is lost. Orthogonality can be restored by inserting estimated values for the missing yields. Paper II gives general methods for obtaining these estimates and discusses the effect of the procedure on the analysis of variance and estimates of error.

Although the estimation of missing values can now best be performed on electronic computers, the sections on the interpretation of the analysis of variance and on the standard errors of treatment comparisons are still relevant. The use of the missing plot technique to investigate apparently anomalous values (last section) is still of importance. The techniques used in the mathematical proofs may also be of general interest.

1969 FRANK YATES

THE ANALYSIS OF REPLICATED EXPERIMENTS WHEN THE FIELD RESULTS ARE INCOMPLETE

F. YATES

(*Rothamsted Experimental Station, Harpenden*)

1. *Introduction.*—The principles of randomization and replication, recently introduced into the design of agricultural field trials, have greatly increased their accuracy, and have rendered possible valid tests of significance and estimates of the experimental errors. But as in all experimental work, it sometimes happens that accidental causes upset the original design, so that the methods of analysis which are ordinarily appropriate require modification. In general, replicated field trials are so arranged that the mean yield of all the plots receiving a given treatment provides the best estimate of the effect of that treatment, free from any extraneous effects, such as fertility differences, which are allowed for in the design. Thus in a randomized block experiment, in which every treatment occurs once in each block, an increase in the fertility of any one block will increase all the treatment means equally, and therefore will not affect comparisons between them. Fertility differences within blocks, on the other hand, will affect comparisons between treatment means, but variation due to this cause also increases the estimate of error and, if the principle of randomization is followed, differences merely due to fertility differences within the blocks will not in general be regarded as providing evidence of real treatment effects. In such an experimental design, where changes in block fertility do not affect differences between treatment means, and conversely changes in treatment effects do not affect differences between block means, blocks and treatments are said to be orthogonal. This property is an extremely valuable one, since not only are the treatment means the best estimates of the treatment effects, but the whole procedure of the analysis of variance, by which an estimate of error may be made, is very simple. When more than two sets of effects are present orthogonality can still be secured, as in a Latin square, where rows, columns, and treatments are all orthogonal to one another, any change in a general fertility of one row, for instance, affecting all the column and treatment means equally. Treatments themselves may also be arranged in orthogonal sets, as when three levels of nitrogen are applied in conjunction with three levels of potash, making nine treatments in all. Such an arrangement provides information not only on the response to nitrogen and to potash separately, but also on the differences in these responses in the presence of different quantities of the other manure. Moreover, orthogonality enables the various effects to be separated in the analysis of variance without difficulty.

If the yields of some plots are lost, or are unreliable, the orthogonality of the original design disappears. Thus in a randomized block experiment with one plot missing, an increase in the fertility of the block containing that plot will affect all treatment means except the one containing the missing plot, and if therefore there is any real difference in the fertility of the different blocks a bias will be introduced into the treatment means, so that their differences are no longer the best estimates of the treatment effects. Nor can the procedure of the analysis of variance appropriate to orthogonal experiments be applied directly to non-orthogonal data. In another paper [1] I have discussed the analysis of non-orthogonal experiments in general, but the case where a few values are missing from an otherwise orthogonal experiment is best dealt with by the special method to be described in this paper.

The problem of analysis may be divided into two parts, the estimation of the magnitude of the effects the experiment was designed to test, and the provision of appropriate tests of significance and estimates of error. The whole problem can be dealt with most successfully by estimating the yields of the missing plots. Such estimates, if properly chosen, when included in the treatment means make these latter efficient estimates of the treatment differences, free from any bias due to other effects such as fertility differences which the experiment was designed to eliminate. After estimating the yields of the missing plots, the ordinary procedure of the analysis of variance suitable to orthogonal experiments may be followed, and though not strictly correct it will be shown that it gives quite satisfactory results in ordinary cases provided the number of degrees of freedom for error is reduced by the number of plots missing. The significance of the results is always slightly exaggerated, though quite negligibly so when only a few values are missing.

The formulae appropriate in the case of a single plot of a randomized block or Latin square have already been given by Miss Allan and Dr. Wishart [2], but no attempt was made to estimate the errors of the treatment differences, and it was assumed that the ordinary procedure of the analysis of variance, if carried out on all the yields, including the estimated yield, was strictly valid when the number of degrees of freedom allotted to error was reduced by one.

The question of the retention or rejection of values which are for some reason considered unreliable may be briefly considered here. In general the mere fact that a value is outstanding does not furnish adequate grounds for its rejection, since it is probable that the causes which produced this outstanding value have also disturbed the other values, though to a lesser extent. The outstanding value must therefore be included in the analysis so that it makes its fair contribution to error. If the magnitude of the error so obtained is so great as to prevent any definite conclusions being drawn, this verdict of non-significance must be accepted. It cannot be too strongly emphasized that the rejection of values, simply because they differ from the rest, is incompatible with any subsequent test of significance and entirely invalidates such tests. On the other hand, if there is sound external evidence for believing a value to be unreliable, such as information that one particular plot and only that plot has suffered from the depredations of some pest, or that previous to the experiment a particular plot was the site of a heap of dung, then rejection may be resorted to. In certain cases when a value is reported to be unreliable owing to some external cause, it is doubtful if the cause is such as to produce any material difference. The experiment itself may legitimately be used to furnish evidence on this point, the reduction in the error sum of squares due to the rejection of the doubtful value being tested for significance against the new error sum of squares.

2. *Method of obtaining missing values.*—The procedure of fitting constants followed by Miss Allan and Dr. Wishart in establishing their formulae becomes involved when applied to complex experiments, and to cases where more than one plot is missing. It was suggested to me by Dr. R. A. Fisher that a much simpler solution might be effected by minimizing the error variance obtained when unknowns are substituted for the missing yields. The validity of this process may be proved rigorously as follows. For simplicity the case of a double classification, e.g. blocks and treatments, is considered, though the proof is perfectly general.

TABLE 1

Treatments.	1	2	3	. .	q
1	y_{11}	y_{12}	y_{13}	. .	y_{1q}
2	y_{21}	y_{22}	y_{23}	. .	y_{2q}
. .	. .	. .	. .	. .	. .
p	y_{p1}	y_{p2}	y_{p3}	. .	y_{pq}

Let Table 1 represent the experimental yields. If these yields are assumed to be made up of additive functions of the blocks and treatments, and an error term, so that

$$y_{rs} = k+t_r+b_s+x_{rs},$$

then the analysis of variance may be regarded as the process of finding the most likely values of the constants k, t_1, t_2,..., t_p, b_1, b_2,..., b_q, and the errors associated with them; that is the values such that $S(x^2_{rs})$ is minimum [3].

In the case where all the yields are known the solution is very simple, being given by

$$\hat{k} = \bar{y}, \quad \hat{t}_r = \frac{1}{q}\sum_1^q y_{rs}, \qquad \hat{b}_s = \frac{1}{p}\sum_1^p y_{rs}. \tag{A}$$

In general, however, it is necessary to minimize the function

$$F = S\,(y_{rs}-k-t_r-b_s)^2, \tag{B}$$

the summation being taken over all existing plot yields.

Now suppose that various plot yields are missing. Assuming for the moment that the most likely values $\hat{k}$, $\hat{t}_1$, $\hat{t}_2$,..., $\hat{b}_1$, $\hat{b}_2$,... of the constants have been found by the above process, put

$$Y_{uv} = \hat{k}+\hat{t}_u+\hat{b}_v \tag{C}$$

for each missing plot. Then Y_{uv} may be taken as an estimate of the yield of the uvth missing plot. If we complete the table of plot yields with these estimates and then perform an ordinary analysis of variance, we shall in fact minimize

$$F' = S(y_{rs}-k-t_r-b_s)^2+S(Y_{uv}-k-t_u-b_v)^2,$$

where the first summation is taken over all the existing plots, and the second over all those of which the yields are missing; this function is the part of the sum of squares allotted to error in the analysis of variance. The second sum of squares is clearly zero by virtue of (C) when the first sum of squares is minimum, i.e. when the most likely values of $k, t_1, t_2, \ldots, b_1, b_2, \ldots$ are obtained. The second sum of squares cannot be negative, and consequently the whole function is now a minimum. But since F' represents the sum of squares over the whole experiment it is minimum for the values of $k, t_1, \ldots, b_1, \ldots,$ given by (A). We may, therefore, instead of minimizing the function F, thus determining directly $\hat{k}, \hat{t}_1, \ldots, \hat{b}_1, \ldots,$ minimize F', using this to determine Y_{uv}, afterwards obtaining $\hat{k}, \hat{t}_1, \ldots, \hat{b}_1, \ldots$ from (A).

3. *General formula for a single missing plot.*—Suppose that there is a multiple classification of the lth order, so that every plot is a member of l classes. Let $p, q, r, \ldots$ be the number of classes in each group, and let n be the number of plots. Thus in a 5×5 Latin square $l = 3$, $p = q = r = 5$, and $n = 25$. Let the sum of all the known yields in the class of the first set of classes containing the missing plot be P, that of the second set be Q, and so on, and let T be the total of all the known yields.

Writing x for the yield of the missing plot and following the ordinary methods of the analysis of variance, the following values for the terms containing x in the sums of squares are obtained:

1st classification . . . $\frac{p}{n}(P+x)^2 - \frac{1}{n}(T+x)^2.$

2nd classification . . . $\frac{q}{n}(Q+x)^2 - \frac{1}{n}(T+x)^2.$

. . . .

Total $x^2 - \frac{1}{n}(T+x)^2.$

Hence the residual (error) sum of squares is

$$x^2 - \frac{p}{n}(P+x)^2 - \frac{q}{n}(Q+x)^2 - \ldots + \frac{l-1}{n}(T+x)^2.$$

Minimizing this, we obtain the equation to determine x,

$$x\{(n+l-1)-(p+q+r+\ldots)\} = (pP+qQ+rR+\ldots)-(l-1)T.$$

In the case of a randomized block experiment, with p treatments, each occurring in q blocks, $n = pq$ and $l = 2$. The sums of the yields of all plots receiving the same treatment and in the same block as the missing plot are P and Q respectively, T being the total yield, and the formula for the missing yield becomes

$$x = \frac{pP+qQ-T}{(p-1)(q-1)}.$$

In the case of a Latin square, with p treatments, $p = q = r$, $n = p^2$, and $l = 3$. If P_r, P_c, and P_t represent the totals of the known yields of the row, column, and treatment from which the plot is missing, the formula becomes

$$x = \frac{p(P_r+P_c+P_t)-2T}{(p-1)(p-2)}.$$

These formulae will be found to agree with those given by Miss Allan and Dr. Wishart [2], save that the plus sign in their formula for a Latin square should be minus.

It should be noted that the general formula deduced above does not include all possible types of classification. In cases where it is not applicable the method of minimizing the error term directly must be followed.

Examples of the application of the formulae for randomized blocks and Latin squares have been given by Miss Allan and Dr. Wishart. There is no need to include any further examples of a single missing plot here, since the use of the formulae is illustrated in the next section in an example where several plots are missing, and again in section 7.

4. *Procedure when several plots are missing.*—In this case the part of the sum of squares allotted to error will be a quadratic function of the yields $x, y, z, \ldots$ of all the unknown plots. On minimizing this function, we shall obtain a set of simultaneous linear equations in $x, y, z, \ldots$.

If P_x is the total of all the known yields in that one of the first set of classes which contains x, P_{yz} the similar total for the class which contains y and z, etc., it being assumed that these two classes contain only these unknown plots and no others, the quadratic function is of the type

$$x^2+y^2+z^2+\ldots-\frac{p}{n}\{(P_x+x)^2+(P_{y2}+y+z)^2+\ldots\}$$

$$-\frac{q}{n}\{(Q_x+x)^2+(Q_y+y)^2+(Q_z+z)^2+\ldots\}-\ldots+\frac{l-1}{n}(T+x+y+z+\ldots)^2.$$

It will be seen that if two unknown plots are members of one class there is a radical difference of form.

The first three linear equations in the case given are

$$x(n+l-1-p-q-\ldots)+y(l-1)+z(l-1)+\ldots$$
$$=pP_x+qQ_x+\ldots-(l-1)T,$$
$$x(l-1)+y(n+l-1-p-q-\ldots)+z(l-1-p)+\ldots$$
$$=pP_{yz}+qQ_y+\ldots-(l-1)T,$$
$$x(l-1)+y(l-1-p)+z(n+l-1-p-q\ldots)+\ldots$$
$$=pP_{yz}+qQ_z+\ldots-(l-1)T.$$

These equations are most easily solved by iterative methods, but in practice there is no need to write them down in the simpler type of experiment, since repeated applications of the formula for a single missing plot, substituting approximate values for all other missing plots, is clearly identical with the ordinary iterative process. The solution converges very rapidly and under ordinary circumstances the second approximation is amply accurate. The details of the numerical calculation are best illustrated by an example.

Example.—Table 2 gives a set of measurements on the intensity of infection of potato tubers inoculated with *Phytophthora Erythroseptica* under various manurial treatments.

TABLE 2

Treat-ments	Blocks 1	2	3	4	5	6	7	8	9	10	Total
O	3·55	2·29	*b*	2·00	3·34	3·83	3·86	3·50	2·23	2·91	27·51+*b*
N	2·30	4·03	2·54	2·82	3·29	2·93	*f*	2·55	2·20	2·30	24·96+*f*
K	3·96	3·62	3·46	2·50	2·94	3·70	3·82	2·54	3·18	3·69	33·41
P	2·99	3·99	2·90	3·97	4·49	4·70	3·86	*h*	3·50	3·59	33·99+*h*
NK	*a*	3·07	3·49	1·07	3·99	3·48	3·80	3·68	3·24	2·70	28·52+*a*
NP	2·36	3·47	2·64	3·17	3·26	3·28	*g*	*i*	3·07	3·12	24·37+*g*+*i*
KP	2·16	2·34	1·96	2·60	3·77	*d*	3·20	3·47	2·67	3·33	25·50+*d*
NKP	3·16	2·52	2·39	3·68	*c*	*e*	3·85	3·36	2·50	4·13	25·59+*c*+*e*
Total	20·48 +*a*	25·33	19·38 +*b*	21·81	25·08 +*c*	21·92 +*d*+*e*	22·39 +*f*+*g*	19·10 +*h*+*i*	22·59	25·77	223·85+*a*+*b* +*c*+*d*+*e*+*f*+*g* +*h*+*i*.

Nine values were missing, indicated by the letters $a, b, c, \ldots, i$. In order to start the process of approximation all the missing values may be assumed to be equal to the mean 3·15. (In cases where the effect of blocks or treatments is very marked it is better to start with the block or treatment means instead of the general mean.) The value of the total, 223·85, must be increased by eight times the value of the mean in order to give an approximate total, 249·05, with only one plot missing. The first approximation for a is then given by

$$(10\times 20{\cdot}48+8\times 28{\cdot}52-249{\cdot}05)/63.$$

This can be very rapidly computed on a calculating machine, multiplying by the reciprocal of 63, 0·01587, in order to avoid division. If a machine is not available a slide-rule will give all necessary accuracy. The value of 2·92 is thus obtained. The value of b may be computed in the same way, without troubling to alter the approximate total 249·05, which may be kept unchanged throughout the first approximation. When obtaining c the treatment total 25·59 must be increased by 3·15 to allow for the other missing plot e. In evaluating e the same treatment total must be increased by the value of c, the value given by the first approximation, 3·67, being taken in preference to the original mean value. The same procedure is followed throughout until a complete set of values is obtained. The new total, 254·10, can now be utilized, being decreased by the first approximation to each plot in turn. The second approximation for a is therefore given by

$$\{10\times 20{\cdot}48+8\times 28{\cdot}52-(254{\cdot}10-2{\cdot}92)\}/63.$$

The complete set of values for both first and second approximations is:

	a	*b*	*c*	*d*	*e*	*f*	*g*	*h*	*i*
1st approximation:	2·92	2·62	3·67	3·26	3·76	3·27	3·61	3·89	3·25
2nd approximation:	2·88	2·58	3·73	3·33	3·76	3·32	3·61	3·89	3·22

The second approximation is accurate to within 0·01. The amended treatment means are found to be:

O	N	P	K	NK	NP	KP	NKP
3·009	2·828	3·341	3·788	3·140	3·120	2·884	3·309

and these provide efficient estimates of the treatment effects, whereas the original treatment means were affected by block differences, the treatment NKP, for example, having too low a mean, 3·199, because of the missing values in blocks 5 and 6, which have high values throughout.

In conclusion, it will be instructive to derive the equations for determining the missing values, in order to illustrate the application of the direct method to numerical data. The derivation of the equations *ab initio* is always the safest procedure in complex experiments where there may be some doubt as to the correct formula for a single missing plot.

In this example the error sum of squares is obtained by subtracting the sums of squares for blocks and treatments from the total sum of squares. Omitting numerical terms, i.e. terms not containing any letter, these sums of squares are as follows:

Total:

$$a^2+b^2+\ldots+i^2-\tfrac{1}{80}(223{\cdot}85+a+b+\ldots+i)^2.$$

Blocks:

$$\tfrac{1}{8}\{(20{\cdot}48+a)^2+(19{\cdot}38+b)^2+\ldots+(19{\cdot}10+h+i)^2\}$$
$$-\tfrac{1}{80}(223{\cdot}85+a+b+\ldots+i)^2.$$

Treatments:

$$\tfrac{1}{10}\{(27{\cdot}51+b)^2+(24{\cdot}96+f)^2+\ldots+(25{\cdot}59+c+e)^2\}$$
$$-\tfrac{1}{80}(223{\cdot}85+a+b+\ldots+i)^2.$$

The equation with d as leading term, for example, can now be written down by differentiating these three quantities with respect to d, and subtracting the last two differentials from the first. On performing this operation, and dividing by 2, we obtain

$$d-\tfrac{1}{8}(21{\cdot}92+d+e)-\tfrac{1}{10}(25{\cdot}50+d)+\tfrac{1}{80}(223{\cdot}85+a+b+\ldots+i)=0.$$

The other equations may be obtained likewise. On multiplication by 80, and simplification, we have, finally,

$$\begin{array}{rrrrrrrrrcl}
+63a & +1b & +1c & +1d & +1e & +1f & +1g & +1h & +1i & = & 209{\cdot}11, \\
+1 & +63 & +1 & +1 & +1 & +1 & +1 & +1 & +1 & = & 190{\cdot}03, \\
+1 & +1 & +63 & +1 & -7 & +1 & +1 & +1 & +1 & = & 231{\cdot}67, \\
+1 & +1 & +1 & +63 & -9 & +1 & +1 & +1 & +1 & = & 199{\cdot}35, \\
+1 & +1 & -7 & -9 & +63 & +1 & +1 & +1 & +1 & = & 200{\cdot}07, \\
+1 & +1 & +1 & +1 & +1 & +63 & -9 & +1 & +1 & = & 199{\cdot}73, \\
+1 & +1 & +1 & +1 & +1 & -9 & +63 & +1 & -7 & = & 195{\cdot}01, \\
+1 & +1 & +1 & +1 & +1 & +1 & +1 & +63 & -9 & = & 239{\cdot}07, \\
+1 & +1 & +1 & +1 & +1 & +1 & -7 & -9 & +63 & = & 162{\cdot}11.
\end{array}$$

The solution of these equations may be effected by iterative methods similar to those used in conjunction with the formula for a single missing plot.

5. *Tests of significance.*—If values are found for the missing plots by the method already explained, and an analysis of variance made on the completed set of yields, the residual sum of squares will be the same as the residual sum of squares obtained when a direct fitting of constants is made, in virtue of the theory given in section 2. The number of degrees of freedom corresponding to this sum of squares will be the number attributable to the residual sum of squares in a similar but complete experiment less the number of missing plots, for the number of fitted constants is the same as in the complete experiment and the

number of independent values (yields) less by the number missing. In so far as the residual variance is concerned, therefore, it is only necessary to reduce the number of degrees of freedom in the analysis of the completed set of values by the number of missing plots.

The variance ascribable to treatments (or any other classification) requires more careful consideration. In general, in non-orthogonal experiments the significance of any set of fitted constants may be tested by finding the further reduction in the sum of squares due to the fitting of these constants simultaneously with constants corresponding to all the other classifications [1]. In certain circumstances it is permissible to neglect some of the classifications (such as treatment interactions) but this point need not concern us here. Therefore, in order to test the significance of treatments in a randomized block experiment with missing plots it is necessary to find the difference of the sum of squares removed by the fitting of constants corresponding to both blocks and treatments and that removed by constants for blocks only.

The sum of squares removed by block and treatment constants can be found by calculating the total sum of squares of the original yields without the missing values (less the correction for the mean) and deducting the residual sum of squares found by analysis of the completed experiment, which, as mentioned above, is equivalent to the residual sum of squares obtained with a direct fitting of the constants. The sum of squares removed by blocks only is obtained directly from the original block totals, following the procedure of the analysis of variance with unequal numbers in the different classes [4, § 44]. The various steps will best be made clear by an example.

Example.—The ordinary analysis of variance of the completed set of values of the experiment on potatoes already described is given in Table 3, and the correct value for the treatment variance is obtained in Table 4. In this latter table the total sum of squares is the sum of the squares of the deviations of all the values in Table 1 from their mean, and the error sum of squares comes from Table 3, giving the blocks and treatments sum of squares; the sum of squares for blocks only comes from the block totals of Table 1, each total after squaring being divided by the number of plots in that block. The difference of these last two quantities gives the proper sum of squares by which treatments may be tested.

The correct mean square for treatments is seen to be less than the mean square obtained by the analysis of the completed set of values. That this is always so can be shown by the following line of reasoning.

TABLE 3. *Analysis of Completed Values*

	Degrees of freedom	*Sum of squares*	*Mean square*	z
Blocks	9	9·7176	1·0797	0·596
Treatments	7	6·5812	0·9402	0·528
Error	54	17·6902	0·3276	
Total	70	33·9890		

TABLE 4. *Analysis of Original Values*

	Degrees of freedom	Sum of squares	Mean square	z
Total . . .	70	32·1012	..	..
Error . . .	54	17·6902	0·3276	..
Blocks and treatments	16	14·4110	..	..
Blocks only . .	9	8·5690	..	..
Difference . .	7	5·8420	0·8346	0·467

For simplicity, the case of a randomized block experiment is chosen, as in section 2.

An analysis of variance may be made with letters for the unknown yields, ignoring the classification to be tested, i.e. in the case of a randomized block experiment an analysis for blocks only. It is clear that an auxiliary set of values may be obtained which will minimize the residual sum of squares of this analysis. In the notation of section 2 the auxiliary set of values is such that

$$S(y_{rs}-k-b_s)^2$$

is as a minimum, the summation being taken over all existing plots, when k and b_s have the values k' and b'_s given by

$$k' = \frac{1}{pq}\sum_{1}^{pq} y_{rs}, \qquad b'_s = \frac{1}{p}\sum_{1}^{p} y_{rs},$$

where the summations now include the missing plots, the auxiliary values being taken.

The analysis of the two completed sets of yields may be set out as follows, the letters denoting the sums of squares.

Ordinary values		*Auxiliary values*	
Blocks	B_o	Blocks	B_a
Treatments	T_o	Residuals	E_a
Error	E_o	Total	S_a
Total	S_o		

If the sum of squares of the deviations of the existing yields from their mean is denoted by S, the reduction in this sum of squares due to fitting block and treatment constants is $S-E_o$, and the reduction due to fitting constants for blocks only is $S-E_a$, in virtue of the relations established in section 2. The difference between these two quantities, which is the correct sum of squares for testing treatments, is therefore E_a-E_o, or

$$T_o-S_o+B_o+S_a-B_a$$

Now S_o-B_o is clearly equal to $S(y_{rs}-\hat{k}-\hat{b}_s)^2$, the summation being taken over all the plots with the ordinary values of the missing plots, and S_a-B_a is equal to $S(y_{rs}-k'-b'_s)^2$, the auxiliary values of the missing plots being taken, so that the terms corresponding to the missing plots are zero in the second summation. The values of k' and b'_s are such that the second expression is a minimum, and therefore $S(y_{rs}-\hat{k}-\hat{b}_s)^2$ must be greater than $S(y_{rs}-k'-b'_s)^2$. Hence the sum of squares due to treatments in the analysis of the completed ordinary set of values (T_o) must be greater than the correct sum of squares by which treatments may be tested. The same argument holds, with the addition of extra sets of constants, for more complex classifications.

In consequence of this result the significance of any effect is always slightly exaggerated in the analysis of variance of the completed set of values, when this analysis is made on the ordinary lines except for the reduction of the number of degrees of freedom for error by the number of missing plots. If, therefore, an effect is found to be not significant in this analysis there is no need to make any further test. On the other hand, if an effect is found to be significant there is theoretically need for further analysis. In practice, however, the difference between the correct and approximate sum of squares is never likely to be great enough to affect the tests of significance seriously, except, perhaps, in cases where a large proportion of the plots is missing.

In the case of a randomized block experiment the amount by which the treatment sum of squares should be reduced is capable of direct expression by a simple formula. The estimated yield of the missing plot of a block containing only one such plot will be denoted by a, the block total excluding this plot by Q_a, and including this plot by V_a. For a block containing two missing plots the corresponding quantities are taken as b_1, b_2, Q_b, and V_b, $\sum_1$, $\sum_2$, etc., denote summation over all blocks containing one, two, or more missing plots. As before p represents the number of treatments.

Omitting terms not containing a, b_1, b_2, etc., we have

$$S_o-B_o = \sum_1 \left\{a^2-\frac{1}{p}(Q_a+a)^2\right\}+\sum_2 \left\{b_1^2+b_2^2-\frac{1}{p}(Q_b+b_1+b_2)^2\right\}+\ldots.$$

Since, as pointed out above, the auxiliary values of the missing plots in the case of a randomized block experiment are merely the block means, the value of S_a-B_a may be immediately obtained by substitution of $Q_a/(p-1)$ for a, $Q_b/(p-2)$ for b_1 and b_2, etc., in the above expression. This gives

$$S_a-B_a = -\sum_1 \frac{1}{p-1}Q_a^2-\sum_2 \frac{1}{p-2}Q_b^2-\sum_3 \frac{1}{p-3}Q_c^2\ldots,$$

Thus

$$S_o-B_o-S_a+B_a = \sum_1 \left\{\frac{p-1}{p}a^2-\frac{2}{p}aQ_a+\frac{1}{p(p-1)}Q_a^2\right\}$$

$$+\sum_2 \left\{b_1^2+b_2^2-\frac{1}{p}(b_1+b_2)^2-\frac{2}{p}(b_1+b_2)Q_b+\frac{2}{p(p-2)}Q_b^2\right\}$$

$$+\sum_3 \left\{c_1^2+c_2^2+c_3^2-\frac{1}{p}(c_1+c_2+c_3)^2-\frac{2}{p}(c_1+c_2+c_3)Q_c+\frac{3}{p(p-3)}Q_c^2\right\}$$

$$+\ldots,$$

and this, by the identity between the n quantities $\alpha, \beta, \gamma, \ldots$

$$(\alpha+\beta+\gamma+\ldots)^2+n(\alpha^2+\beta^2+\gamma^2+\ldots) \equiv (\alpha-\beta)^2+(\alpha-\gamma)^2+(\beta-\gamma)^2+\ldots,$$

and the substitution of V_a for Q_a+a, etc., reduces to

$$\frac{1}{p(p-1)}\sum_1 (V_a-pa)^2$$

$$+\frac{1}{2p(p-2)}\sum_2 \{2V_b-p(b_1+b_2)\}^2+\frac{1}{3p(p-3)}\sum_3 \{3V_c-p(c_1+c_2+c_3)\}^2+\ldots$$

$$+\tfrac{1}{2}\sum_2 (b_1-b_2)^2+\tfrac{1}{3}\sum_3 \{(c_1-c_2)^2+(c_1-c_3)^2+(c_2-c_3)^2\}+\ldots,$$

which is the expression required. This formula clearly shows that the difference is never likely to be large.

Applying the formula to the potato experiment already worked out, the difference is found to be 0·7391, which agrees with the result previously obtained. The difference for the sum of squares due to blocks in the same experiment is 1·6630, giving a corrected mean square of 0·8950 ($z = 0{\cdot}503$). Neither of these corrections seriously alters the significance of the results. If only a single plot is missing the differences are likely to be quite trivial.

The strict analysis of more complex classifications is not so simple, for the reason that the reduction in the sum of squares due to the fitting of all sets of constants other than the one under test cannot be directly computed. In a Latin square, for example, in order to test the significance of the treatments it is necessary to compute the reduction in the sum of squares due to fitting constants for rows and columns only; this can best be done by finding an auxiliary set of values for the missing plots, neglecting the treatment classification, i.e. by utilizing the formula for randomized blocks, taking rows and columns in place of blocks and treatments. For a Latin square the difference between the sums of squares does not lend itself to simple expression by means of a formula except in the case when one plot is missing. In the notation of section 2 the difference then is

$$\frac{1}{(p-1)^2(p-2)^2}\{(p-1)P_t+P_r+P_c-T\}^2.$$

Similar methods could be applied to other complex classifications such as occur when it is desired to split up the general treatment effect into different components, but in practice the analysis of the completed yields will be sufficiently accurate, except in cases where a large proportion of the results are missing. In such cases the direct fitting of constants may prove the easier method of approach.

In the example already given seven degrees of freedom for treatments may be split up in the analysis of the completed set of yields into seven single degrees of freedom, corresponding to direct effects and first and second order interactions, but it is unnecessary to proceed farther with the analysis here, which is better made in conjunction with the other part of the experiment (not reproduced).

6. *Errors of treatment means.*—If the main tests of significance are made by means of an analysis of variance there is no need to make any very precise determination of the standard errors of the treatment means. Nevertheless the provision of standard errors is useful as an indication of the accuracy of the results, and it will therefore be of interest to see what differences are made by missing plots.

When only a single plot is missing the treatment mean containing the estimated value of the missing plot can easily be expressed as a linear function of all the known yields, whence its variance can be calculated in the ordinary manner. If σ^2 is the variance of a single yield, the variance in the case of a randomized block is, in the previous notation,

$$\frac{1}{q}\left\{1+\frac{p}{(p-1)(q-1)}\right\}\sigma^2,$$

and in the case of a Latin square it is

$$\frac{1}{p}\left\{1+\frac{p}{(p-1)(p-2)}\right\}\sigma^2.$$

The variances of the other treatment means are, of course, σ^2/q and σ^2/p respectively. Thus in a randomized block experiment with six treatments and four replications the variance is 1·40 times what it would have been with no plot missing, and in a 5×5 Latin square 1·42 times. The variance of the differences of this and any other mean is therefore in both these cases about 1·2 times what it should have been. In other words the significant difference is about 10 per cent. greater.

When only one plot is missing the treatment mean containing this plot is uncorrelated with the other treatment means, and therefore the variance of the difference of two means is the sum of the variances of these means. If more than one plot is missing this is in general only true of differences between means, one of which contains no missing plot. Consequently, in order to find the variance of the difference of two treatment means both of which contain missing plots, it is necessary to express the difference of the two means as a linear function of the known yields. If the number of missing plots is greater than two or three this involves heavy algebra, which is not worth while. An alternative method of approach by fitting constants directly and inverting the determinant is equally laborious.

Although when a number of plots are missing the computation of the standard errors of the differences of treatment means is not practicable, in the case of a randomized block it is easy to fix upper and lower limits between which the errors must lie. A lower limit is provided by ignoring the block classification altogether, and an upper limit by rejecting all those blocks which do not contain both treatments. Thus in the experiment already considered the variance of the difference between the treatment means for NP and NK must be greater than $(\frac{1}{8}+\frac{1}{9})\sigma^2$ and less than $\frac{2}{7}\sigma^2$, i.e. it must lie between $0{\cdot}235\sigma^2$ and $0{\cdot}286\sigma^2$. A good working rule is to give half weight to each plot of a treatment which has no corresponding plot in the same block belonging the other treatment. This will give $(\frac{1}{7\cdot 5}+\frac{1}{8})\sigma^2$, or $0{\cdot}258\sigma^2$ as the value of the variance in question.

A similar rule might be formulated for a Latin square, each yield which has the corresponding yield of the other treatment missing in either row or column being given a weight of $\frac{2}{3}$, or $\frac{1}{3}$ if both the corresponding yields are missing.

7. *Use of the missing plot technique in testing interactions.*—In an ordinary analysis of variance it sometimes happens that when a set of interactions is found to be significant, there is a probability that the whole of this significance may be accounted for by a single outstanding value. This hypothesis may be rigorously tested by omitting the outstanding value, and supplying a new value in its place by means of the formula for a missing plot. The new variance for interaction (the number of degrees of freedom being diminished by one) can then be calculated and tested. The procedure is best illustrated by an example.

Example.—Table 5 gives the logarithms of the mean yields of four varieties of cotton at seven different centres. At each centre a 4×4 Latin square was laid out. Logarithmic yields are taken in order to equalize the variances at different centres, the standard errors being found to be roughly proportional to the mean yields. The estimate of the variance of a single entry of the table from the variances at the different centres was found to be 0·0008686.

TABLE 5

Centres	*A*	*B*	*C*	*D*	*Total*	$D-\frac{1}{3}(A+B+C)$
1	0·190	0·097	0·149	0·246	0·682	0·101
2	0·528	0·538	0·450	0·593	2·109	0·088
3	0·290	0·276	0·238	0·356	1·160	0·088
4	1·021	0·959	0·981	1·248	4·209	0·261
5	0·164	0·176	0·149	0·233	0·722	0·070
6	0·566	0·546	0·494	0·627	2·233	0·092
7	0·650	0·614	0·601	0·647	2·512	0·025
Total	3·409	3·206	3·062	3·950	13·627	..

The analysis of variance is given in Table 6. Varieties, places, and the interactions between them are all significant. The last column of Table 5, giving the difference between *D* and the mean of the other three varieties, indicates that the extra high yield of *D* at the high yielding centre, 4, is exceptional, and may therefore account for most of the interaction.

TABLE 6

	Degrees of freedom	*Sum of squares*	*Mean square*	*z*
Varieties . .	3	0·064897	0·021632	..
Places . .	6	2·316032	0·386005	..
Interactions .	18	0·032653	0·001814	0·3680
Total . .	27	2·413582	..	..
Error . .	42	..	0·0008686	..

The value of the yield of D given by the missing plot formula is 1·064. The new analysis of variance based on this value is given in Table 7.

TABLE 7

	Degrees of freedom	Sum of squares	Mean square
Varieties . .	3	0·039965	0·013322
Places . .	6	2·115157	0·352526
Interactions .	17	0·010941	0·000644
Total . .	26	2·166063	

The interaction mean square is now below expectation, and the single abnormal value therefore accounts entirely for the original significance. The sums of squares due to varieties and places are also reduced in the new analysis. This is to be expected since the average varietal and place effects excluding the anomalous value of D at 4 are now what are approximately represented. Tests of significance may be made if so desired. The difference between the original and new interaction sums of squares, 0·021712 (one degree of freedom) also gives a valid test of significance which may be employed to test a single outstanding value when interactions as a whole are not significant, if there is *a priori* ground for believing that this value may in fact be different from the others.

Summary.—The procedure introduced by Miss Allan and Dr. Wishart for supplying a missing value in a table of experimental results, such as the plot yields of a field trial, so that the treatment means form unbiased and efficient estimates of the treatment effects, is here extended to enable any number of missing values to be replaced, it being shown that the method of derivation adopted previously is equivalent to the simpler method of minimizing the error term in the ordinary analysis of variance. The solution of a complex example is effected by iterative methods.

The validity of analysis of variance on the completed table of values is investigated. It is shown that when the degrees of freedom allotted to error are reduced by the number of values replaced, there is little disturbance, provided that the number of missing values is not too great. Such disturbance as there is always exaggerates the significance of the results. The standard errors of the treatment means are also briefly discussed.

The use of the missing-plot technique for further analysing interactions, whose significance is believed to be due to a few anomalous values, is illustrated by the analysis of a set of varietal trials on cotton.

REFERENCES

1. F. YATES, The Principles of Orthogonality and Confounding in Replicated Experiments. J. Agric. Sci., 1933, **23**, Pt. 1, 108–145.
2. F. E. ALLAN and J. WISHART, A Method of Estimating the Yield of a Missing Plot in Field Experimental Work. J. Agric. Sci., 1930, **20**, Pt. 3, 399–406.
3. J. O. IRWIN, Mathematical Theorems involved in the Analysis of Variance. J. Roy. Stat. Soc., 1931, **94**, 284–300.
4. R. A. FISHER, Statistical Methods for Research Workers. Edinburgh: Oliver and Boyd, 1925 (4th ed. 1932).

PAPER III

THE FORMATION OF LATIN SQUARES FOR USE IN FIELD EXPERIMENTS

FROM THE EMPIRE JOURNAL
OF EXPERIMENTAL AGRICULTURE
VOLUME I, No. 3, pp. 235–244, 1933

Author's Note

When the principle of randomization was first applied to Latin square designs, Fisher considered that the ideal procedure was to select at random (with equal probability) one from all possible Latin squares of the required size. This paper was written with this dictum in mind. Its immediate object was to provide agriculturists with a means of making such a selection for squares of up to size 6 × 6, which had by then been enumerated. The main continuing interest of the paper is the discussion on the sub-sets of all possible squares that give an unbiased estimate of error and the effect of correlations of neighbouring plot yields on the z-distribution.

At the time it was written it was recognized that the Knut-Vik (knight's move) squares might be expected to give results of greater than average precision, but with a correspondingly inflated estimate of experimental error, thus leading to too few significant results. Conversely, for the diagonal squares the reverse was the case. The deliberate exclusion of such squares from those selected was, however, at that time regarded as heretical. Formal justification for this course came much later with the introduction of the idea of restricted randomization (see Paper XII).

1969 FRANK YATES

THE FORMATION OF LATIN SQUARES FOR USE IN FIELD EXPERIMENTS

F. YATES

(*Rothamsted Experimental Station*)

1. *Introduction*

REPLICATION was originally introduced into agricultural field trials with the object of securing greater accuracy by distributing the different treatments more evenly over the field and so balancing out the fertility differences. The estimation of the error to which the results were subject did not at first receive much consideration, but with the evolution of a logical basis for inductive reasoning it became apparent that wholly systematic arrangements of plots in field trials were incapable of furnishing valid estimates of error, without which it was impossible to draw any completely objective conclusions from the results. In order to obtain a valid estimate of error it is essential that there shall be some element of randomness in the arrangement of the plots, and that any restrictions which are imposed for the purpose of eliminating soil heterogeneity shall be such that their effects on the reduction of error can be clearly isolated by the procedure of the analysis of variance. The simplest type of arrangement fulfilling these conditions is that of randomized blocks, where the plots are arranged so that every treatment occurs an equal number of times in every block, the random element in the arrangement being very simply introduced by assignment of the treatments to the different plots within each block entirely at random.

For many experiments randomized block arrangements are eminently suitable, being very flexible and capable of a wide variety of applications. On the other hand the Latin square, where each treatment occurs once and once only in each row and each column, though by no means so flexible, may be expected on the average to eliminate soil heterogeneity more completely, at any rate when the size of the square is not too large, since soil variations in two directions at right angles affect all treatments equally. Such elimination is peculiarly attractive in agriculture because most agricultural operations are carried out in long strips, and by arranging that these strips shall be parallel to one of the sides of the square variation from this source is eliminated.

The value, as a means of eliminating fertility differences, of square arrangements of plots, satisfying the conditions of the Latin square, was early recognized. When first introduced, however, the importance of an unbiased estimate of error was not realized, and the arrangements adopted were all systematic, usually of some specially simple type, or alternatively of a type which was believed to be capable of removing most completely the soil differences ordinarily existing. The term Latin square, used by Euler [1] in his study of the enumeration of the different possible square arrangements, was introduced into agricultural science to designate a square chosen at random out of the totality of these arrangements.

Striking practical confirmation of the failure of systematic square arrangements to give a valid estimate of error was afforded by the work of Tedin [2]. He took twelve 5×5 Latin-square arrangements, four of them systematic, and applied them to data from eight uniformity trials collected from various sources, which provided in all 91 separate squares of 25 plots. The treatment and error sums of squares were computed for each of the arrangements. Since the treatments were dummy the error sum of squares (12 degrees of freedom) should on the average be three-quarters of the total sum of squares after eliminating rows and columns (16 degrees of freedom), and significant divergence from this fraction would be an indication of a biased estimate of error. Two of the arrangements chosen were knight's move patterns, and these gave an average fraction of 0·7720, significantly higher than 0·75, indicating that such arrangements are in general more accurate than the average of all random arrangements, but appear to be less accurate. This accords with the claims advanced for this type of design. The actual amount of the bias introduced may best be indicated by the number of times the result may be expected to be judged significant on a 5 per cent. basis, this, Tedin concludes, being 4·2 times out of 100 in the squares under consideration. The data are not extensive enough, however, for this value to be considered as at all accurately determined.

The other pair of systematic arrangements were of the diagonal pattern (the second square of Fig. 2 and the same square rotated through a right angle). These arrangements are chosen as an example for discussion in the next section. The particular type of fertility correlation there considered does not appear to give rise to any particular bias in the error, though there may be an excess of very high and low values. Tedin, however, reasoning on the proximity or dispersion of similarly treated plots, concludes that this pair of arrangements should have a low average estimate of error. The actual average value of the fraction of the sum of squares allotted to error, 0·7291, supports his conclusion, though the difference from 0·75 is not fully significant. There is also some slight evidence of an excess of very low values in one of the arrangements.

In contradistinction to the two systematic arrangements tested, the distribution of z in the eight Latin squares (728 values) compared excellently with expectation.

2. *Conditions for an Unbiased Estimate of Error*

The element of randomization may be introduced into Latin-square arrangements by making a selection of one square from a whole set of squares. It remains to consider what particular types of set will give an estimate of error which is unbiased when averaged over all the squares of the set.

The z distribution, on which the tests of significance depend, is established in the first place on the assumption that the yields are an uncorrelated sample from a normally distributed population. (The word sample here implies the whole set of yields that go to form a single experiment.) It requires an infinity of experiments to generate the distribution of the statistic z.

The procedure of the analysis of variance enables correlations between whole rows and columns to be eliminated. The z distribution will still hold, whatever the Latin-square arrangement adopted, if the residuals (i.e. the remainders after deducting from each plot the general mean and the deviations from the general mean of the means of the row and column in which it occurs) form a sample from a normal population without any correlations except those introduced by the conditions that the sum of the residuals of each row and each column is necessarily zero.

Actually the assumption of no correlation between the residuals of neighbouring plots is unjustified. A complex and unknown system of correlations must be assumed, these correlations being what is left of the inherent correlations between neighbouring plots after the removal of row and column effects. As an example we may consider a system of experiments in which the fertility of the land is distributed in strips running diagonally across the lay-out of the square. In this case there will be a positive correlation between the residuals of the plots lying on the system of parallels crossing the square in this direction, i.e. between the plots bearing the same number in Fig. 1. In the limiting case, when

1	2	3	4	5	6
2	3	4	5	6	7
3	4	5	6	7	8
4	5	6	7	8	9
5	6	7	8	9	10
6	7	8	9	10	11

FIG. 1.

A	B	C	D	E	F
B	C	D	E	F	A
C	D	E	F	A	B
D	E	F	A	B	C
E	F	A	B	C	D
F	A	B	C	D	E

FIG. 2.

the correlations are perfect, the residuals of all plots bearing the same number will be equal in every sample, and since the sums of the residuals of every row and column are zero the residuals of plots 1 will equal the residuals of plots 7, &c.

If in this example a Latin square of the pattern shown in Fig. 2 be chosen as the experimental lay-out in the series of experiments, it is clear that the treatment sum of squares will tend to be too large and the error sum of squares too small, and in the limiting case where the correlation is perfect, the error sum of squares will always be zero and consequently z will always be infinite. Conversely, if the arrangement chosen be that of Fig. 2, but turned through a right angle, the treatment sum of squares will tend to be too small, and in the limiting case it will always be zero, and z will be negatively infinite.

This example provides a simple illustration of how, given a certain system of correlations between the residuals, the use of a single Latin square will give estimates of error which are biased. If, however, a suitably chosen set of Latin squares is used, such bias in the estimate of error can be entirely eliminated.

The simplest set of squares which will eliminate the bias in the estimate of error, whatever the system of correlation between the residuals, is that obtained by permuting all the rows except the first of any chosen square in all possible ways. In such a set of squares every pair of plots not in the same row or column receives like treatments equally frequently.

In the set of squares generated from that shown in Fig. 2, for example, A of the second column will occupy each of the last five places in the column 24 times, so that the first plot of the first column has like treatment to each of the other plots of the second column 24 times. The second plot of the first column will receive treatment B 24 times and will then have like treatment with the first plot of the second column. It will receive treatment C 24 times, and of these 24 times C will fall in the last four places of the second column 6 times each. The same occurs with D, E, and F, and therefore this plot is associated with each of the last four plots of the second column 24 times.

The absence of bias in the error can be shown as follows. Let the residuals of a single $n \times n$ experiment, after correcting for rows and columns, be given by Table 1.

TABLE 1

$$\begin{array}{cccccc} x_{11} & x_{12} & x_{13} & \cdot & \cdot & x_{1n} \\ x_{21} & x_{22} & x_{23} & \cdot & \cdot & x_{2n} \\ x_{31} & x_{32} & x_{33} & \cdot & \cdot & x_{3n} \\ \cdot & \cdot & \cdot & & & \cdot \\ x_{n1} & x_{n2} & x_{n3} & \cdot & \cdot & x_{nn} \end{array}$$

The sums of the residuals of every row and every column are necessarily zero. If the totals of the residuals for the n treatments are T_A, T_B, T_C, ... T_K, then the sum of squares due to treatments is

$$\frac{1}{n}(T_A{}^2+T_B{}^2+T_C{}^2+\ldots+T_K{}^2).$$

On permuting the last $n-1$ rows in all the $(n-1)!$ possible ways, $(n-1)!$ such treatment sums of squares will be obtained. Since every pair of plots not in the same row and same column have like treatment the same number of times, and the square of each treatment total when multiplied out contains n squares of residuals and $\frac{1}{2}n\,(n-1)$ pairs of products, it follows that the total of all the $(n-1)!$ treatment sums of squares must reduce to

$$\frac{(n-1)!}{n}\cdot Sx_{rs}{}^2+\frac{(n-1)!}{n(n-1)}Sx_{rs}x_{tu}, \qquad t\neq r,\ u\neq s,$$

where the first summation represents the sum of all squares of residuals and the second summation twice the sum of the products of residuals of all pairs of plots not in the same row or column.

The sum of the products of any residual, say x_{11}, with all the residuals not in the same row or column, is equal to the square of that residual, for $x_{22}+x_{23}+\ldots+x_{2n}=-x_{21}$, &c., so that the required sum is equal to $-x_{11}\,(x_{21}+x_{31}+\ldots+x_{n1})$, i.e. $x_{11}{}^2$. Hence the last summation in the above expression is equal to $Sx_{rs}{}^2$, and therefore the mean treatment sum of squares reduces to

$$\frac{1}{n-1}Sx_{rs}{}^2.$$

This corresponds to $n-1$ degrees of freedom, whereas the total sum of squares of the residuals after correcting for rows and columns, Sx_{rs}^2, corresponds to $(n-1)^2$ degrees of freedom. There is consequently no bias in the treatment sum of squares if the whole set of Latin squares is applied to the results of a single experiment. There can therefore equally be no bias if the whole set of squares is applied to each of the whole population of experiments, or, since this population is infinite, if a single square selected at random from the set is applied to each member of the whole population.

The permutation of the last $n-1$ rows removes all bias from the estimate of error, but since the first row is the same for all squares, individual treatments are still not free from bias if a single experiment is considered, or from correlation if we consider the whole population of experiments. Such bias and correlation can be eliminated by permuting all the letters of the square. Alternatively, the elimination can be performed just as effectively by permuting all the columns.

Although the employment of such a method of randomization will eliminate the bias due to error and any correlation between treatments, it appears that with a given system of correlations between the residuals, the z distribution will not be exactly realized. Since the greatest bias in z is introduced when the pattern of the square coincides most nearly with the correlation pattern, or cuts across it most completely, as in the example discussed above, it would seem theoretically preferable to choose a square at random from all the possible squares of given size, since with such a choice the maximum bias will occur less frequently than if the square is always chosen from a set as defined above, which happens to contain a square coinciding most nearly with the correlation pattern. The point is largely theoretical, for in agricultural experiments the correlation system of an infinity of experiments can hardly be very large, or of the type that will coincide at all completely with any possible Latin square.

Instead of considering a whole population of experiments, we may consider the z distribution which would be obtained if a set of arrangements be applied to a single experiment. Such a distribution is necessarily discontinuous, for the number of different values of z is some fraction of the number of arrangements in the set. In order that the distribution shall approximate to a continuous distribution the number of arrangements included in the set must be large. This is the basis of the test made by Eden and Yates [3] on 256 height measurements of wheat, which displayed marked departure from the normal form of distribution. The data were arranged in eight blocks of four values (each value being the sum of eight measurements) and a sample of 1,000 arrangements taken at random from the whole $(4!)^8$ possible arrangements, the z being computed for each arrangement. There are $(4!)^7$ discrete values of z, and it is shown that the sample of 1,000 from this discontinuous population conforms satisfactorily with the theoretical z distribution for an infinite population of normally distributed data.

In order that such a distribution shall approximate to a continuous distribution the number of arrangements included in the set must be large, and this may be regarded as a further point in favour of using a set of squares as large as conveniently possible. In the case of 5×5 squares, for example, the permutation of all the rows, except the first, of a given square will only give twenty-four different values of z, whereas the utilization of all squares will give 1,344 such values.

All possible squares up to 6×6 have been enumerated and are easily presentable [4]; they are therefore illustrated in the next section. A single square of each size from 7×7 to 12×12, is also given. From these typical squares experimental arrangements may be derived as required.

3. *Typical Squares*

The most extensive set of squares that can be easily derived from a single square is that generated by the permutation of all rows, all columns and all letters. This type of permutation, which we have styled a *transformation* in the enumeration of the 6×6 squares [4], is the basis of the presentation of the 5×5 and 6×6 squares. The greatest number of $n \times n$ squares that can possibly belong to any transformation set is clearly $(n!)^3$, since the rows, columns, and letters may all be permuted independently in $n!$ ways (including no change), but all transformations of a given square do not necessarily give different squares, so that the actual number in the set may be very much less than this. In the case of 6×6 squares, for example, the greatest number of squares in any one set is $\frac{1}{4}(6!)^3$ and the least number $\frac{1}{216}(6!)^3$.

Although all transformations do not in general give different squares, every square of the set (including the original square) must be generated an equal number of times, when all the $(n!)^3$ transformations (including no change) are made. Thus in the case of 6×6 squares every square of a set containing $\frac{1}{4}(6!)^3$ squares must be generated four times. It follows that if a transformation be chosen at random from all the possible $(n!)^3$ transformations, this transformation, when applied to any given square of a set, will generate a square which is a random selection from all possible squares in the set. In order to make a random selection from the $(n!)^3$ transformations it is only necessary to choose some new random order for the rows and columns and a random substitution for the letters (by numbered cards, balls, &c., or a table of random numbers [5]).

The fraction $1/n!\,(n-1)!$ of all the $n \times n$ Latin squares will be what are called reduced squares, i.e. squares with the first row and first column in the given order A.B.C.D.... Each transformation set contains the same fraction of reduced squares. From each reduced square $n!\,(n-1)!$ squares can be generated by permuting all the rows except the first, and all the columns, or alternatively, all the rows except the first, and all the letters, and either of these generating processes, when applied to all reduced squares, will generate all possible squares.

The 3×3 squares are all included in a single transformation set, which contains one reduced square, so that there are $3!\,2!$ or twelve 3×3 squares in all. The reduced square is illustrated in Fig. 3. A random selection from all the squares may be made by permuting at random the last two rows and all the columns or letters.

A	B	C
B	C	A
C	A	B

FIG. 3. The 3×3 squares.

The 4×4 squares form two transformation sets, one containing three reduced squares, all of which are illustrated in Fig. 4 (*a*), and the other a single reduced square, Fig. 4 (*b*). A random selection from all the squares may be made by selecting one of the reduced squares at random and permuting the last three rows and all the columns or letters.

A B C D	A B C D	A B C D	A B C D
B A D C	B C D A	B D A C	B A D C
C D B A	C D A B	C A D B	C D A B
D C A B	D A B C	D C B A	D C B A
(*a*)	(*a*)	(*a*)	(*b*)

FIG. 4. The 4×4 squares.

The 5×5 squares also form two transformation sets, one containing 50 and the other 6 reduced squares. Space does not permit us to set out all these 56 reduced squares in full, and a single example of each transformation set is therefore given, in Fig. 5, from which all squares of the set may be generated by the permutation of all rows, all columns, and all letters.

To make a random selection from all possible 5×5 squares with an equal probability of obtaining any one square, it is first necessary to select one or other of the transformation sets, in such a manner that the probability of selection is proportional to the number of squares in the set. This may be done most simply by selecting a number at random from the numbers 1–56, and choosing the first set when a number between 1 and 50 is obtained, and the second when a number between 51 and 56 is obtained, as indicated by the key numbers printed below the squares.

A B C D E	A B C D E
B A D E C	B C D E A
C E A B D	C D E A B
D C E A B	D E A B C
E D B C A	E A B C D
1–50	51–6

FIG. 5. The 5×5 squares.

The 6×6 squares, Fig. 6, form 22 transformation sets, comprising in all 9,408 reduced squares. Ten of these sets form 5 pairs of sets, such that all squares of one set of the pair are the conjugates of the squares in the other set. (One square is said to be the conjugate of another if the rows of one square, taken in order, correspond to the columns of the other, also taken in order.) These pairs of sets are illustrated by a single example each, but are distinguished from those sets which include conjugates by being given two sets of key numbers. Since all the squares of one of a conjugate pair of sets, when rotated through a right angle, give the squares of the other set, the agriculturist can hardly be concerned with the distinction between a square and its conjugate, for the orientation of an experiment is itself usually arbitrary. The purist may satisfy himself by performing the rotation if a number included in the lower group of key numbers is obtained.

Apart from the occurrence of these conjugate pairs of sets, the method of selection of a square at random from all the possible 9,408. 6! 5! 6×6 squares is identical with that already described in the case of the 5×5 squares.

I	II	III
A B C D E F B C F A D E C F B E A D D E A B F C E A D F C B F D E C B A	A B C D E F B C F E A D C F B A D E D E A B F C E A D F C B F D E C B A	A B C D E F B A F E C D C F B A D E D C E B F A E D A F B C F E D C A B
0001–1080 1081–2160	2161–3240	3241–4320

IV	V	VI
A B C D E F B A E F C D C F B A D E D E A B F C E D F C B A F C D E A B	A B C D E F B A E C F D C F B A D E D E F B C A E D A F B C F C D E A B	A B C D E F B A F E C D C F B A D E D E A B F C E C D F B A F D E C A B
4321–5400	5401–5940 5941–6480	6481–7020

VII	VIII	IX
A B C D E F B C D E F A C E A F B D D F B A C E E D F B A C F A E C D B	A B C D E F B A E F C D C F A E D B D C B A F E E D F C B A F E D B A C	A B C D E F B A E F C D C F A B D E D E B A F C E D F C B A F C D E A B
7021–7560	7561–7920 7921–8280	8281–8640

X	XI	XII
A B C D E F B C F A D E C F B E A D D A E B F C E D A F C B F E D C B A	A B C D E F B C A F D E C A B E F D D F E B A C E D F C B A F E D A C B	A B C D E F B C A E F D C A B F D E D E F B A C E F D A C B F D E C B A
8641–8820	8821–8940 8941–9060	9061–9180

XIII

A B C D E F
B C A F D E
C A B E F D
D F E B A C
E D F A C B
F E D C B A

9181–9240

XIV

A B C D E F
B C A E F D
C A B F D E
D F E B A C
E D F C B A
F E D A C B

9241–9280

XV

A B C D E F
B A F E D C
C D A B F E
D F E A C B
E C B F A D
F E D C B A

9281–9316
9317–9352

XVI

A B C D E F
B A E C F D
C E A F D B
D C F A B E
E F D B A C
F D B E C A

9353–9388

XVII

A B C D E F
B C A F D E
C A B E F D
D E F A B C
E F D C A B
F D E B C A

9389–9408

FIG. 6. The 6×6 Latin squares.

No enumeration has as yet been made of squares larger than 6×6. In Fig. 7 we give six squares, with sides from 7 to 12, from which any square of the transformation sets which contain them may be generated by the permutation of all rows, columns, and letters amongst themselves. These transformation sets, or even the smaller sets generated by the permutation of rows and columns, or either and letters, will give sets of squares amply large enough to serve all agricultural purposes.

A B C D E F G
B D E F A G C
C G F E B A D
D E A B G C F
E C B G F D A
F A G C D E B
G F D A C B E

7×7

A B C D E F G H
B C A E F D H G
C A D G H E F B
D F G C A H B E
E H B F G C A D
F D H A B G E C
G E F H C B D A
H G E B D A C F

8×8

A B C D E F G H I
B C E G D I F A H
C D F A H G I E B
D H A B F E C I G
E G B I C H D F A
F I H E B D A G C
G F I C A B H D E
H E G F I A B C D
I A D H G C E B F

9×9

A B C D E F G H I J
B G A E H C F I J D
C H J G F B E A D I
D A G I J E C B F H
E F H J I G A D B C
F E B C D I J G H A
G I F B A D H J C E
H C I F G J D E A B
I J D A C H B F E G
J D E H B A I C G F

10×10

A B C D E F G H I J K
B A J I D C F K H G E
C K H A B I J F D E G
D C G J I K E B F A H
E J B G K H D C A I F
F E I C G A K J B H D
G F D B H J A I E K C
H I K F A D B E G C J
I D E H J B C G K F A
J G A K F E H D C B I
K H F E C G I A J D B

11×11

A B C D E F G H I J K L
B L G C D J K E H A F I
C K A B F L I D G H J E
D F I A L E C G J B H K
E D F G J K A L C I B H
F H K E G C D B A L I J
G I D F K H J A L C E B
H E L J C A B I K D G F
I J B L H G F K D E A C
J C E K A I H F B G L D
K G J H I B L C E F D A
L A H I B D E J F K C G

12×12

FIG. 7.

4. *Summary*

1. The conditions which must be fulfilled in selecting Latin-square arrangements for agricultural field trials, if an unbiased estimate of error is to be obtained, are discussed.

2. Examples of squares up to size 12×12 are given, from which experimental arrangements may be derived by simple processes of permutation. All squares up to size 6×6 have been enumerated elsewhere, and the totalities of these squares are presented here in compact form.

REFERENCES

1. L. EULER, Recherches sur une nouvelle espece de quarres magiques. Verh. v. h. Zeeuwsch Genootsch. der Wetensch. Vlissingen, 1782, **9**, 85–239.
2. O. TEDIN, The Influence of Systematic Plot Arrangement upon the Estimate of Error in Field Experiments. J. Agric. Sci., 1931, **21**, 191–208.
3. T. EDEN and F. YATES, On the Validity of Fisher's z Test when applied to an Actual Example of Non-normal Data. J. Agric. Sci., 1933, **23**, 6–17.
4. R.A. FISHER and F. YATES. The 6 × 6 Latin Squares. Proc. Camb. phil. Soc., 1934, **30**, 492–507.
5. L.H.C. TIPPETT, Tracts for Computers. No. XV. Random sampling numbers. Cambridge University Press.

Paper IV

COMPLEX EXPERIMENTS

From THE SUPPLEMENT TO THE JOURNAL
OF THE ROYAL STATISTICAL SOCIETY
Volume II, No. 2, pp. 181–247, 1935

Author's Note

Paper IV was read before the Industrial and Agricultural Research Section—which later became the Research Section—of the Royal Statistical Society. It was the first formal exposition to a general statistical audience of factorial design and the use of confounding in such designs. At the time it was written there was still strong opposition to the use of factorial designs, and much of the paper is devoted to arguing the case for them. It contains several ideas not previously published, including the estimation of error from high order interactions, balanced confounding in $3 \times 2 \times 2$ and similar designs, and balanced incomplete blocks. The methods exemplified in it were later expanded for systematic use by agriculturists in *Technical Communication No.* 35 of the Commonwealth Bureau of Soils.

An interesting feature of the paper is the estimation of the efficiency of various types of design from actual experimental results: previously it had been considered that this could only be done by using uniformity trial data.

Confounded designs had been criticized from time to time, without any attempt to determine whether responses varied appreciably from block to block, on the ground that such differential responses would vitiate tests of significance of certain interactions. This point is also examined in the paper: no evidence of any such differential responses was found.

It is the custom of the Royal Statistical Society to publish the discussion on a presented paper. I would have liked to reproduce the whole of this, but for reasons of space have included only my written reply. For the full discussion the Society's Journal must be consulted.

1969 FRANK YATES

COMPLEX EXPERIMENTS

By F. YATES, M.A.

[Read before the Industrial and Agricultural Research Section of the Royal Statistical Society, May 23rd, 1935, SIR WILLIAM DAMPIER, Sc.D., F.R.S., in the Chair.]

1. *Introduction.*

SEVERAL papers have recently been read before this Section or published in the Supplement to the *Journal* which have dealt with one aspect or another of the randomized block and Latin square methods of carrying out replicated experiments.

These methods were first developed in connection with field trials in agriculture, but since their inception it has been abundantly clear that they are of very wide application, and are therefore of interest to experimental workers in almost all branches of science and technology. The paper I propose to give to-night may be considered as belonging to the same series. In it I intend to deal with another aspect of the methods which has not yet been discussed, namely the part which treatments play in experimental design.

Following the previous writers, I shall describe the special technique which has been developed in agricultural field trials, but it is hoped that the paper will prove of interest not only to agricultural workers, but also to workers in many other branches of research. For while the complex method has been most fully developed in connection with agriculture, the special difficulties occasioned by the necessity of eliminating soil heterogeneity both limit its application and complicate the method. It would appear that its use in other fields in which the material is more homogeneous would be even more fruitful.

In the absence of specialized knowledge of the problems and conditions which occur in other branches of research and industry it is impossible to give detailed examples of the utility of complex design in these fields. But a few general suggestions may be made to show the wide scope of the method. For the rest I would stress that there are many more ways than one of carrying out experimental work, and only with a knowledge of the basic principles of the available methods, and a good acquaintance with the sources of variation of the experimental material, can the experimenter hope to arrive at an efficient technique. For this reason it has appeared to me profitable to discuss somewhat fully the application of these methods to agricultural research. The difficulties to be overcome in other fields will not be identical, but many of them will be similar, and in so far as this is the case the solutions which have been arrived at in agriculture will be of material assistance. In so far as they differ fundamentally, new technique will be required, but such technique can only be evolved when in actual contact with the difficulties themselves.

As examples of particular applications of complex experimentation (or, as we shall call it, *factorial design*) to fields other than agriculture, I may first instance research into cotton-spinning problems carried out at the Shirley Institute. In some experiments described by Tippett in a paper recently read before this Section,[6] the adoption of complex methods enabled him to investigate three or four factors at once instead of one. His experimental problem was peculiarly simple, in that he was prepared to assume the virtual independence of the several factors he was investigating, so that the effect of variation in one factor could be regarded as substantially the same whatever the values of the other factors. (This method is described in Section 9.) In biological experiments such assumptions cannot be regarded as satisfactory, but I imagine that in many experiments on manufacturing processes they are more justified.

Similar problems might occur in almost any branch of manufacture. In testing different types of motor tyre, for instance, we might wish to vary both the speed and the load, or both and the surface; or in actual road tests types of car, drivers, average speeds and similar factors might be varied. Factorial design would result in considerably more efficient utilization of experimental resources.

Then there is the wide field of biological enquiry, of which the work I am about to describe is a special branch. I need only here instance dietetic experiments on animals, and especially on human beings (where experimental material is hard to come by). Here the interactions between the various components of any diet are of vital importance, and factorial designs would appear to be as necessary as they are in the very similar problems encountered in agronomic research.

As regards the applications of the designs alluded to in Section 10 (for which I shall propose the name of *incomplete randomized blocks*) I need only mention three instances. In human beings the resemblance between monozygotic twins is well known; if this is to be utilized experimentally we are limited to the equivalent of blocks of two. In certain virus experiments which involve the inoculation of leaves of young plants only five suitable leaves are growing on the plants at one time, so that blocks must be limited to five leaves each. And in many laboratory determinations in which variation in conditions seriously affects the results the number of determinations which can be carried out simultaneously (or approximately simultaneously) is strictly limited.

It has been thought better to retain throughout the terminology of agricultural field experiments, rather than create a more generalized terminology which might be applicable to all experimental material. This course recommends itself the more in that workers in any field will in practice refer to their experimental units by their appropriate names, so that some transposition of terms when passing from one field to another will always be necessary.

Since the whole evolution of experimental technique is bound up with questions of efficiency I have considered it worth while to devote a certain amount of space to a discussion of the actual gain in efficiency that is likely to result in agricultural field trials by the use of factorial design. This part of the paper will be of particular interest to agriculturists, since objections against factorial design have been made on the ground of actual loss of efficiency. But here again the discussion is likely to be of general interest, as indicating the lines to be followed in similar investigations in other fields.

I am afraid that certain parts of the paper will prove difficult reading to those not familiar with the principles of the analysis of variance. This is inevitable if undue bulk and tedious repetition are to be avoided. To those not acquainted with this branch of statistical technique I would recommend that they make a preliminary study of Fisher [1], and Fisher and Wishart [3], before they attempt to follow the analytical procedure of the examples given in this paper. I would also like to emphasize that a full appreciation of all the points of the various designs is more likely to be obtained if the actual examples given are worked over in full on a computing machine. It is to make this possible that I have reproduced here the numerical values of the individual plot yields.

Finally, it may be well here to emphasize two further points in connection with the evolution of an efficient experimental technique, which are sometimes lost sight of. The first is that the amount of work it is profitable to expend in developing a technique depends on the amount of experimental work of a given type that is likely to be undertaken. It is because the need for agricultural experimentation is so widespread and persistent that it has been worth while devoting considerable research merely to improving agricultural experimental technique. The other is that it is not necessarily any reflection on the ability of an experimenter if the methods he has employed and advocated are later found to be less efficient than other newer methods. In many fields of research there is a tendency for a sort of vested interest to grow up round an experimental method, which leads to its defence on entirely illogical and unreasonable grounds, and to a very grudging acceptance of newer methods.

2. *An Experimental Problem.*

Suppose that a new plant of agricultural importance is introduced into a country. (An example has recently been provided by the introduction of sugar beet into England.) What is the right way to set about determining the best varieties and the appropriate manurings and cultivations?

One procedure, extensively practised, is to divide the problem into a large number of smaller problems and attack each one separately. One set of experiments will be started to determine the best variety, a second set to determine the best manuring, a third the best cultivations. Nor need, nor does, the subdivision stop there. Responses to the three standard fertilizers, nitrogen, phosphate and potash, for instance, may be relegated to separate experiments.

This procedure has on the face of it a deceptive appearance of simplicity. The questions formulated are themselves simple:—Is one variety better than another? Is the yield increased by the application of nitrogen? Their answers can be obtained with an apparently high precision. But there is one very cogent objection. Clearly the experimenter on fertilizers, who, we will imagine, decides to confine his enquiries at the start to response to nitrogen, must choose some variety on which to experiment. He will probably choose what he considers is the best variety. After three years of experiment the experimenter on varieties may announce that some other variety is markedly superior. Are all the experiments on fertilizers now worthless, in that they apply to a variety that will no longer be grown? The experimenter on fertilizers will probably answer that the response of the two varieties is not likely to be widely different, and that his conclusions therefore still hold. But he has no experimental proof of this, only his experience of other crops, and would not have even this last if he and all other experimenters on fertilizers had persistently only experimented on one variety.

If the experimenter on varieties is so rash as to criticize his results on these grounds, however, and has himself laid down some standard of manuring for his varietal trials, the experimenter on fertilizers can effectively turn the tables by pointing out that the varietal trials are also of little value, being carried out at a level of manuring different from what he proposes to recommend.

Had the enquiries been combined into one system of experiments, so that all varieties were tested in conjunction with all levels of nitrogen, this imaginary controversy could not have arisen; for definite information would be obtained on whether all varieties did, in fact, respond equally to nitrogen. Moreover, if it was found that they did, a considerable gain in efficiency on the primary questions would result (provided that the experimental errors per plot were not greatly increased), since each plot would enter into both primary comparisons and would thus be used twice over. If, on the other hand, differential response was demonstrated, then although the response of the chosen variety would be known with less accuracy than if the whole experiment had been carried out with that variety, yet the experimenter might count himself lucky in that the possibility of false conclusions due to using another variety in his fertilizer trials had been avoided. Moreover, the conclusion as to the best variety might also require modification in the light of the differences in response to fertilizer.

When new varieties are being selected this greatly understates the advantage, for the essential and valuable difference of one variety over another may lie just in its ability to respond to heavy dressings of fertilizer; at the customary levels of manuring, it may be, the yields are about the same. Nor need this response be direct. In the case of wheat, for instance, the limit of nitrogenous manuring is determined less by what the plant can make use of than by what it can stand up to without lodging.

In practice it will seldom be possible to include every variety it is desired to test in the fertilizer trials. This, however, is no argument against including a representative selection of varieties. If no substantial differences in fertilizer response are discovered with such a selection we may then, reasoning inductively, conclude that it is improbable that substantial differences exist for *any* variety.

The method of experimentation in which two or more sets of treatments, or treatments and varieties, are taken in all combinations, was originally called complex experimentation, but inasmuch as this term may be taken to imply complexities of other kinds, I propose, following Fisher, to use the terms *factorial design* and *factorial experiments*.

3. *History.*

The idea of using all combinations of various sets of treatments in fertilizer experiments, and not only those which appear advisable on the grounds of the particular theory held at the moment, is a very old one. The two contrasting methods are well exemplified in classical fertilizer experiments on wheat and barley at Rothamsted.

The wheat experiment on Broadbalk was first laid down in the season 1843–4, but was first put on a permanent basis in 1851–2. Table I gives the table of treatments, the majority of which have been continued without change ever since. It will be seen that the effect of any particular mineral salt is for the most part given by

TABLE I.

Broadbalk Permanent Wheat.

Scheme of Manuring adopted in 1852.

Plot. No.	Manuring.				
2*b*	Farmyard manure				
3–4, 2*a* (half), 20	No manure				
5	—	P	K	Na	Mg
6	½N	P	K	Na	Mg
7, 15*a*	N	P	K	Na	Mg
8	1½N	P	K	Na	Mg
16	2N	P	K	Na	Mg
10	N	—	—	—	—
11	N	P	—	—	—
13	N	P	K	—	—
12	N	P	—	3⅔Na	—
14	N	P	—	—	2¾Mg
2*a* (half)	—	—	K	Na	Mg
17 alternating	N	—	—	—	—
18 alternating	—	P	K	Na	Mg
9	Nitrate of Soda, etc.				
15*b*, 19	Rape Cake, etc.				

N = Ammonium salts.
P = Superphosphate.
K = Sulphate of potash.
Na = Sulphate of soda.
Mg = Sulphate of magnesia.

the difference of only a single pair of plots. On the other hand, the effect of increasing dressings of nitrogen is well determined by five levels of nitrogenous manuring. In plan the experiment consisted of a series of long, narrow plots (349 yards × 6·90 yards), each ½ acre in area, stretching the full length of the field. The numbering was consecutive from 2 to 19 across the field. A few of the plots were divided longitudinally into two halves (indicated by *a* and *b*). 3–4 was one plot. 2*a* was divided into two halves transversely, and 20 was one-third the length of the field.

The design of the Hoosfield barley experiment, laid out in 1852, is shown in Fig. 1. The mineral treatments were in strips along the field, and the nitrogen treatments in strips at right angles to them, so that all combinations of

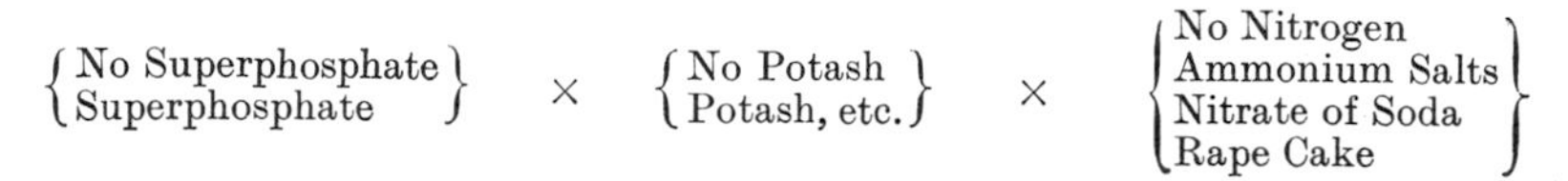

were included, as well as farmyard manure and a few miscellaneous treatments.

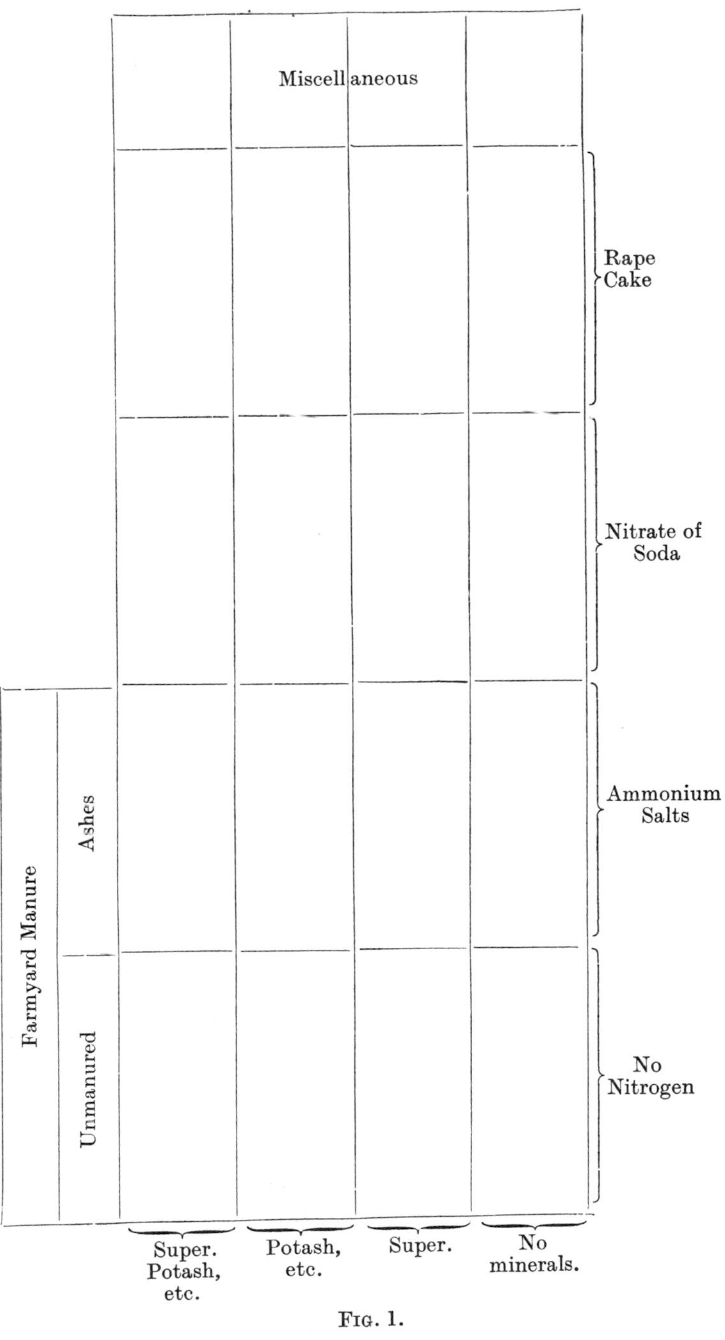

FIG. 1.

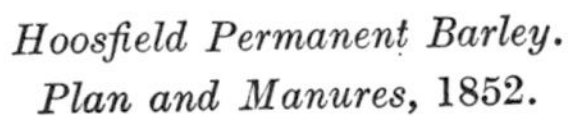

Hoosfield Permanent Barley.
Plan and Manures, 1852.

It will be seen that almost the whole of the field enters into the contrast between potash and no potash, and superphosphate and no superphosphate, and into the response to the various forms of nitrogen. Moreover, information is provided on the differences in response to any one of the fertilizers in the presence and absence of the others. Thus in one respect the design is markedly superior to Broadbalk. Unfortunately, however, the effect of broad fertility irregularities is likely to be serious. If, for instance, there is a fertility gradient from bottom to top of the field, even with no real effect of nitrogen a response to nitrogen would be indicated, rape cake being apparently the best. The fact that all four rape cake plots did better than all four no-nitrogen plots, which if the plots were arranged at random might be taken as evidence of a real difference, could here be equally regarded as evidence of fertility differences. In Broadbalk, on the other hand, fertility differences in one direction (actually up and down the slope) are entirely eliminated, since the plots stretch the whole length of the field. A certain amount of internal replication is also accidentally provided if it is assumed (as seems likely) that the effects of sulphate of soda and sulphate of magnesia are small.

The realization of the large and unknown errors introduced by fertility differences led to the use of replication, and also to a tendency to confine attention to a few simple comparisons, so that greater accuracy might be attained. Fertilizer experiments, for instance, tended to a simplified Broadbalk system of treatments, such as

(1), *n*, *np*, *npk*,

or (1), *np*, *nk*, *pk*, *npk*,

the particular choice of treatments depending largely on the particular theory of fertilizer response held by the experimenter. (Throughout the paper the separate treatments will be indicated by small letters and the treatment effects and their interactions by capital letters. No treatment is indicated by (1), or by the suffix 0. The use of (1) instead of 0 for no treatment enables the expressions for the main effects and interactions to be written down by the rules of algebra.)

The development of complex experimentation in modern agricultural research is primarily due to Fisher. As early as 1926 [2] he made a very strong recommendation in favour of complex experiments, which it will be of interest to recall.

> "In most experiments involving manuring or cultural treatment, the comparisons involving single factors, *e.g.* with or without phosphate, are of far higher interest and practical

importance than the much more numerous possible comparisons involving several factors. This circumstance, through a process of reasoning which can best be illustrated by a practical example, leads to the remarkable consequence that large and complex experiments have a much higher efficiency than simple ones. No aphorism is more frequently repeated in connection with field trials than that we must ask Nature few questions, or, ideally, one question, at a time. The writer is convinced that this view is wholly mistaken. Nature, he suggests, will best respond to a logical and carefully thought out questionnaire; indeed, if we ask her a single question, she will often refuse to answer until some other topic has been discussed."

Since this date complex experimentation based on factorial design has been extensively practised at Rothamsted, and to a lesser extent elsewhere. The methods have evolved gradually, particularly in the direction of confounding, and with growing familiarity has come increased confidence. In the following sections I propose to give an outline of the methods at present in use. It is manifestly impossible, within the scope of a single paper, to enter into explanations of the detailed mechanism of all the devices. Detailed examples of three arrangements have, however, been included as exemplifying most of the points at issue.

4. *The* $2 \times 2 \times 2 \times \ldots$ *System of Treatments.*

The simplest type of factorial design is that in which there are only two treatments in each set, so that the total number of combinations is some power of 2. If we ignore the different forms of nitrogen the system of treatments in the Hoosfield barley experiment already referred to is of this type.

Let us first consider the case where there are two treatments, say nitrogen (n) and potash (k) only. We then have the four treatment combinations

$$(1),\ n,\ k,\ nk.$$

In the experiment on peas referred to in detail later the total yields on the six replicates of these treatments (ignoring slag, which in fact produced no effect) were respectively

(1)	n	k	nk
317·3,	365·1,	307·5,	327·1.

There are three independent comparisons, represented in the analysis of variance by three degrees of freedom. Clearly the comparisons can be made in many ways. We might, for instance, consider the response over the six replicates to n alone, given by

$n - (1)$, here 47·8, and to k alone, given by $k - (1)$, here $-$ 9·8, and the increase or decrease over the sum of these responses when n and k are applied in combination, namely $(nk - (1)) - (n - (1)) - (k - (1))$, or $nk - n - k + (1)$, here $-$ 28·2.

The difference between nk and k, however, also furnishes information on the response to n, being the response to n in the presence of k, and if the response to n and k are, in fact, independent, this response is equal to the response to n in the absence of k, except for experimental errors. A general measure of the effect of n is therefore provided by the mean of the responses to n in the presence and absence of k. The appropriate estimate of this is

$$N = \tfrac{1}{2}\{(nk - k) + (n - (1))\} = \tfrac{1}{2}(n - (1))\,(k + (1))$$
$$= \tfrac{1}{2}(19{\cdot}6 + 47{\cdot}8) = \tfrac{1}{2}(67{\cdot}4) = 33{\cdot}7,$$

and this may be defined as the *main effect* of n.

Equally the mean response to k in the presence and absence of n, *i.e.* the main effect of k, is estimated from

$$K = \tfrac{1}{2}\{(nk - n) + (k - (1))\} = \tfrac{1}{2}(n + (1))\,(k - (1))$$
$$= \tfrac{1}{2}(-38{\cdot}0 - 9{\cdot}8) = \tfrac{1}{2}(-47{\cdot}8) = -23{\cdot}9.$$

A measure of the lack of independence of n and k is given by the difference of the responses to n in the presence of k and in the absence of k, namely $(nk - k) - (n - (1))$, or what is the same thing, the difference of the responses to k in the presence and absence of n, namely $(nk - n) - (k - (1))$. One half of either of these is equal to

$$N \times K = \tfrac{1}{2}\{nk - n - k + (1)\} = \tfrac{1}{2}(n - (1))\,(k - (1))$$
$$= \tfrac{1}{2}(19{\cdot}6 - 47{\cdot}8) = \tfrac{1}{2}(-38{\cdot}0 + 9{\cdot}8) = \tfrac{1}{2}(-28{\cdot}2) = -14{\cdot}1.$$

This quantity is called the *interaction* between n and k, and represents one-half the differential response of n in the presence and absence of k.

The conventional introduction * of the factor $\tfrac{1}{2}$ has two advantages: first, the main effects and the interaction are all determined with equal precision, and secondly, the differences between individual treatment combinations can be determined by the direct addition or subtraction of the appropriate main effects and interactions. Thus the response to n in the absence of k is given by

$$N - (N \times K) = 33{\cdot}7 - (-14{\cdot}1) = 47{\cdot}8,$$

and the response to both fertilizers in combination, $nk - (1)$, is given by

$$N + K = 33{\cdot}7 - 23{\cdot}9 = 9{\cdot}8.$$

* This was made since the meeting.

It will be noted that this last response is not affected by the interaction.

It should be noted that the expressions for the main effects and interactions are really a matter of definition, the interactions being measures of the departure of the observed differences from the law implied in the definition of the main effects. Here the main effects are so defined as to imply an additive law between the effect of n and the effect of k; this is statistically convenient, and in agriculture appears to provide a good representation of the type of effects usually observed. But it should be clearly understood that the additive law has been provisionally imposed by the statistician and is not implicit in the data.

All the above responses are expressed in terms of the effects on totals of six plot yields. In practice the final presentation of the results will be in terms of some such units as hundredweight per acre, involving some further conversion factor. In the computation the factor $\frac{1}{2}$ will be combined with this conversion factor. Thus in the above example, where each plot has an area of 1/70th acre, and the yields are in pounds, the main effects and interaction will be found by multiplying 67·4, — 47·8, and — 28·2 (each of which is composed of the sums and differences of 24 plot yields) by

$$\frac{70}{12 \times 112}$$

to give the responses and interaction in hundredweight per acre.

It may be noted that the quantities 67·4, — 47·8, and — 28·2 are the differences of the marginal totals and the cross difference of the two-way table :

	(1).	n.	Total.
(1)	317·3	365·1	682·4
k	307·5	327·1	634·6
Total	624·8	692·2	1317·0

The sums of squares in the analysis of variance corresponding to these three quantities can be found by squaring the three numbers 67·4, — 47·8, and — 28·2 and dividing each square by 24, since each is the difference of two sums of 12 plots each. (The appropriate divisor of the square of any linear function of the plot yields

$$l_1 y_1 + l_2 y_2 + \ldots.$$

representing a single degree of freedom is $l_1^2 + l_2^2 + \ldots.$ The same rule holds for finding the divisor of the sum of squares of deviations of quantities whose differences represent a set of two or more degrees of freedom, provided that no plot yield occurs in more than one quantity.)

The subdivision of the treatment sum of squares is given in Table II.

TABLE II.

	D.F.	Sum of Squares.	Mean Square.
Nitrogen	1	189·28	189·28
Potash	1	95·20	95·20
Interaction	1	33·14	33·14
Total	3	317·62	105·87

These three sums of squares possess an interesting statistical property, in that they add up to the total sum of squares (three degrees of freedom) for treatments. This is an indication that the three estimates are statistically independent. Technically the three linear functions corresponding to the three single degrees of freedom are said to be *orthogonal.* For a full discussion of the practical implications of orthogonality and non-orthogonality the reader is referred to Yates [8]. Two linear functions of the plot yields

$$l_1y_1 + l_2y_2 + \dots$$
$$l_1'y_1 + l_2'y_2 + \dots$$

are orthogonal if

$$l_1l_1' + l_2l_2' + \dots = 0.$$

Although the appropriate estimates of the treatment effects are not always orthogonal, this is usually the case in a well-designed experiment, a fact which introduces a certain logical completeness into the analysis of variance. The additive property of sums of squares corresponding to orthogonal degrees of freedom is extensively used in the practical calculations required in the analysis of variance. It is important, therefore, that workers should be able to recognize the existence of non-orthogonality so as not to assume the additive property when it does not hold, an assumption which will lead to very serious errors.

The first two of the comparisons given in the third paragraph of this section are not orthogonal. The linear functions are

$$y_n + y_n' + y_n'' + \dots - y_0 - y_0' - y_0'' - \dots$$
$$y_k + y_k' + y_k'' + \dots - y_0 - y_0' - y_0'' - \dots$$

and the sum of the products ll' is $+ 6$. The sums of squares are

$\frac{1}{12}\{n - (1)\}^2$	190·40
$\frac{1}{12}\{k - (1)\}^2$	8·00
$\frac{1}{24}\{nk - n - k + (1)\}^2$	33·14
Total	231·54

The total is very different from the total sum of squares for treatments, 317·62. But each of these sums of squares may be legitimately compared with the error mean square by means of the z test, if it is desired to test the effect in question, just as the quantities $n - (1) = 47{\cdot}8$, etc. may be tested by the t test, which is the exact equivalent of the z test for a single degree of freedom.

The error mean square in this experiment (twelve degrees of freedom) is 15·44 (Table VI), so that the mean effects of both nitrogen and potash (Table II) are both significant (potash giving a depression), but the interaction is not significant.

It is, of course, possible that an interaction of some kind exists although it is not significant. It will be noticed that the response to n is only significant in the absence of k, and the depression with k is only significant in the presence of n. But in the absence of consistent interactions of this type over a series of experiments and of any knowledge of when they are likely to occur, the average response to n, 33·7, and the average depression with k, $-$ 23·9, are the best estimates to adopt in assessing the advantages of these fertilizers.

It will also be noticed that should the interaction exist and the yields of the individual combinations of treatments be substantially correct, the experimenter who divided his experiment into two parts, one on n with a basal dressing of k, and one on k with a basal dressing of n, would certainly discover the inadvisability of applying k, but would also come to the conclusion that the response to n was quite trivial, a seriously erroneous judgment.

In general if there is no evidence of interaction the mean responses to the two factors may be taken as the appropriate measures of the responses to these factors, which may be regarded as additive. If interaction exists, then usually information will be required on the responses to each factor in the presence and absence of the other. The experiment furnishes information on this point, but with only half the precision (*i.e.* with $\sqrt{2}$ times the standard error) of the mean comparisons. Nothing is lost, however, as even should the results be judged insufficiently precise, the information already obtained can be combined with that added later.

In certain cases it may be judged as a result of the first series of experiments, that more precise information will only be required on certain of the comparisons embodied in the factorial design. One variety, for instance, might prove itself so superior to all others that interest in response to fertilizers centred on the response of that variety. But the fact which may have emerged that some other variety did respond considerably more, although of no immediate practical interest, might still be of considerable value to the plant breeder in his task of combining the desirable features of several varieties.

The 2×2 system may be easily extended to any power of 2. We need only consider the case of a $2 \times 2 \times 2$ system. If the treatments are the three standard fertilizers n, p and k in all combinations, the main effect of n may be defined as the mean of the responses to n in the presence of all combinations of the other two fertilizers, and will then be estimated from

$$\begin{aligned} N &= \tfrac{1}{4}\{(n - (1)) + (np - p) + (nk - k) + (npk - pk)\} \\ &= \tfrac{1}{4}(n - (1))\,(p + (1))\,(k + (1)). \end{aligned}$$

The first order interactions of n and p may be defined as the mean of the interactions of n and p in the presence and in the absence of k, and will then be estimated from

$$\begin{aligned} N \times P &= \tfrac{1}{2}\{\tfrac{1}{2}(np - n - p + (1)) + \tfrac{1}{2}(npk - nk - pk + k)\} \\ &= \tfrac{1}{4}(n - (1))\,(p - (1))\,(k + (1)). \end{aligned}$$

The expressions for the main effects of p and of k, and the first order interactions of n and k and of p and k, will be similar. There is still one degree of freedom unaccounted for. This represents what is called the *second order interaction* between n, p and k, which is one half the difference between the interactions of n and p in the presence and absence of k, or between the interactions of n and k in the presence and absence of p, or between the interactions of p and k in the presence and absence of n, all these being equivalent and equal to

$$\begin{aligned} N \times P \times K &= \tfrac{1}{2}\{\tfrac{1}{2}(npk - nk - pk + k) - \tfrac{1}{2}(np - n - p + (1))\} \\ &= \tfrac{1}{4}(n - (1))\,(p - (1))\,(k - (1)). \end{aligned}$$

A numerical example of this type of subdivision is provided by the experiment on peas at Biggleswade, further discussed in Section 7.

The expressions tabulated on page 195 for various typical responses may be noted.

If the second order interaction is ignored the response to all three factors in conjunction is equal to the sum of the main effects of the three factors.

Response to:	Expression in terms of	
	treatment combinations.	main effects and interactions.
n (p absent, mean of k and no k)	$\frac{1}{2}\{nk + n - k - (1)\}$	$N - (N \times P)$
n (p and k absent)	$n - (1)$	$N - (N \times P) - (N \times K) + (N \times P \times K)$
n and p (mean of k and no k) ...	$\frac{1}{2}\{npk + np - k - (1)\}$	$N + P$
n and p (k absent)	$np - (1)$	$N + P - (N \times K) - (P \times K)$
n, p and k (complete fertilizer) ...	$npk - (1)$	$N + P + K + (N \times P \times K)$

5. *Subdivision of Sets of Degrees of Freedom.*

Very frequently we wish to introduce sets of treatments which contain more than two treatments each. In experiments involving fertilizers, for instance, it may be desirable to investigate more than one level of each fertilizer, or several different forms of the same factor, as in the nitrogen treatments of Hoosfield barley. In combined varietal and manurial trials it is usually advisable to include at least three varieties.

The primary subdivision of the treatment degrees of freedom follows the same lines as in $2 \times 2 \times 2 \times \ldots$ experiments, but each main effect and interaction will now contain more than a single degree of freedom.

The actual structure is best made clear by an example. Table III

TABLE III.

Oats Variety and Manuring Experiment.

Treatment Totals.

	n_0.	n_1.	n_2.	n_3.	Total.
v_1	429	538	665	711	2,343
v_2	480	591	688	749	2,508
v_3	520	651	703	761	2,635
Total... ...	1,429	1,780	2,056	2,221	7,486

gives the treatment totals (six replicates) of an experiment on oats involving three varieties and four levels of nitrogen. (The design of this experiment, which was a split-plot one, will be discussed in detail later.)

The partition of the treatment degrees of freedom and sum of squares is as follows:

	D.F.	Sum of Squares.	Mean Square.
Varieties	2	1786·36	893·18
Nitrogen	3	20020·50	6673·50
Interaction	6	321·75	53·63
Treatments... ...	11	22128·61	—

The sum of squares due to varieties is calculated from the varietal marginal totals, and that due to nitrogen from the nitrogen totals, while the interaction is usually obtained by subtraction of these two sums of squares from the total sum of squares due to treatments. The computation is exactly analogous to the computation of the sums of squares for blocks, treatments and error in a randomized block experiment. In fact " error " in such experiments is formally composed of the interactions of blocks and treatments.

This, however, does not exhaust the possibilities of subdivision. Any set of n degrees of freedom can be divided into n single degrees of freedom in an infinity of ways. Only the division which coincides with whatever physical facts it is desired to bring into prominence will normally be of interest, but in certain cases some especially simple formal division is useful for purposes of confounding.

In our example the three degrees of freedom for nitrogen may naturally be divided into the one degree of freedom representing the linear component of the response curve, and a second representing the quadratic component, and a third the cubic.

The dressings represent equal increments of nitrogen and the linear term is therefore (Fisher [1], Section 27) given by some fraction of

$$-3S(n_0) - S(n_1) + S(n_2) + 3S(n_3),$$

the quadratic term by

$$S(n_0) - S(n_1) - S(n_2) + S(n_3),$$

and the cubic term by

$$-S(n_0) + 3S(n_1) - 3S(n_2) + S(n_3).$$

The squares of these quantities, divided by 360, 72 and 360 respectively, will give the sums of squares attributable to each degree of freedom. These are

	D.F.	Sum of Squares.
Linear term	1	19536·4
Quadratic term	1	480·5
Cubic term	1	3·6
Total	3	20020·5

Most of the variation due to nitrogen response is accounted for by the linear term. The appropriate error mean square is 177·08, and the quadratic term is therefore not significant. The cubic term is below expectation.

The interactions between varieties and nitrogen may be similarly split up by calculating regression terms for each variety and taking the sums of the squares of the deviations between them. Indeed in testing for differential response this should be done, since such differential response is most likely to reveal itself in differences between the regressions for the different varieties; the whole set of interactions contains four degrees of freedom which are likely in any case to be small when there is little curvature in the average response curve.

In this particular example the total sum of squares for the six degrees of freedom is only 321·75, so that no interaction can possibly be significant, but the calculation may be performed as a formal exercise. The numerical values of the regression terms are

	Linear.	Quadratic.	Cubic.
v_1	973	63	99
v_2	904	50	22
v_3	775	73	−85

The sums of the squares of the deviations of these quantities divided by 120, 24 and 120 respectively give the following sums of squares :

	D.F.	Sum of Squares.	Mean Square.
$N_1 \times$ varieties	2	168·35	84·18
$N_2 \times$ varieties	2	11·08	5·54
$N_3 \times$ varieties	2	142·32	71·16
Total	6	321·75	—

If it were desired to test some theoretical response curve, the appropriate division of the sum of squares would be into the part accounted for by the curve, with as many degrees of freedom as there were arbitrary constants in the curve, and a part representing deviations from the curve. The test of the adequacy of the theory would be the non-significance of the latter response.

6. *Split Plot Arrangements.*

As already mentioned, the chief practical difficulty in the application of factorial design to agricultural field experiments is the fact that the number of treatment combinations rapidly becomes large with increasing complexity, with resultant large blocks and inadequate elimination of fertility differences. A further practical difficulty is that many agricultural operations cannot be conveniently carried out on small plots.

To meet these difficulties various modifications of the randomized block and Latin square lay-outs have been devised. The two of chief interest are the splitting of plots for subsidiary treatments, and the confounding of high-order interactions. We will first consider the split-plot type of design.

In principle this is very simple. The whole-plot treatments are arranged in the ordinary manner in a randomized block or Latin square, and each plot is subdivided into two or more parts to which are assigned at random the two or more sub-treatments. The analysis is also simple, especially if the division is only into two. In this case all that is necessary is to perform two separate analyses, with separate errors, one on the totals of the whole plots, and the

v_3	n_3 156	n_2 118	n_2 109	n_3 99	v_3
	n_1 140	n_0 105	n_0 63	n_1 70	
v_1	n_0 111	n_1 130	n_0 80	n_2 94	v_2
	n_3 174	n_2 157	n_3 126	n_1 82	
v_2	n_0 117	n_1 114	n_1 90	n_2 100	v_1
	n_2 161	n_3 141	n_3 116	n_0 62	

v_3	n_2 104	n_0 70	n_3 96	n_0 60	v_2
	n_1 89	n_3 117	n_2 89	n_1 102	
v_1	n_3 122	n_0 74	n_2 112	n_3 86	v_1
	n_1 89	n_2 81	n_0 68	n_1 64	
v_2	n_1 103	n_0 64	n_2 132	n_3 124	v_3
	n_2 132	n_3 133	n_1 129	n_0 89	

v_2	n_1 108	n_2 126	n_2 118	n_0 53	v_1
	n_3 149	n_0 70	n_3 113	n_1 74	
v_3	n_3 144	n_1 124	n_3 104	n_2 86	v_2
	n_2 121	n_0 96	n_0 89	n_1 82	
v_1	n_0 61	n_3 100	n_0 97	n_1 99	v_3
	n_1 91	n_2 97	n_2 119	n_3 121	

← rows →

Area of each plot : 1/80 acre. (28·4 links × 44 link rows.)

FIG. 2.

Oats Variety and Manuring Experiment.

Plan and yields in quarter lb.

other on the differences of the pairs of sub-plots. This latter analysis will contain an extra degree of freedom representing the direct effect of the sub-plot treatment. The interactions of the sub-plot treatment with the whole plot treatments will also appear in this analysis. If the subdivision is into more than two parts, the sums of squares corresponding to the components of the second analysis are usually obtained by subtraction, in the same manner as the interaction sum of squares in Section 5.

The oats variety and manurial experiment already referred to furnishes a good example of this type of lay-out. The varieties were sown in six randomized blocks of three plots each, and each plot was subdivided into four for the four levels of nitrogen.

The plan and yields of grain are shown in Fig. 2. The full analysis of variance is given in Table IV. The analysis is on a sub-plot basis, sums of squares from the whole-plot totals being divided by an extra 4, since each is the total of four sub-plots.

TABLE IV.

Oats Variety and Manuring Experiment.

Analysis of Variance (Sub-Plot Basis).

		D.F.	Sum of Squares.	Mean Square.
Whole Plots	Blocks	5	15875·28	3175·06
	Varieties	2	1786·36	893·18
	Error	10	6013·30	601·33
	Total	17	23674·94	—
Sub-Plots	Nitrogen	3	20020·50	6673·50
	N × Varieties... ...	6	321·75	53·63
	Error	45	7968·76	177·08
	Total	71	51985·95	—

In general in such experiments the comparisons between the sub-plot treatments are likely to be more accurately determined than those between the whole-plot treatments, since they depend on comparisons between closely adjacent plots. This is the case here, where the two error mean squares are 177·08 and 601·33, or in the ratio of 1 to 3·40. (The reader should satisfy himself that the ratio between these mean squares does represent the comparative accuracy of means containing the same number of sub-plots but subject to sub-plot and whole-plot errors respectively.) By replacing each treatment mean square by the corresponding error mean square we obtain the equivalent of a uniformity trial. Combining the two resultant error sums of squares then gives an estimate of what the error mean square would have been had all treatments combinations been randomized. This gives

$$\frac{12 \times 601{\cdot}33 + 54 \times 177{\cdot}08}{66} = 254{\cdot}22$$

subject to errors of estimation. Thus the accuracy of the varietal comparison would have been considerably increased, with some, but not a proportionate, loss of accuracy on the responses to nitrogen and their interactions with varieties. On the varietal comparisons we obtain 2·37 (=601·33/254·22) times the original information, and on the comparisons involving nitrogen 0·70 (= 177·08/254·22) times the original information.

In general, therefore, if the main effects of all treatments are required with equal accuracy, split-plot arrangements are not to be recommended where the whole-plot treatments are arranged in randomized blocks, except in the case in which whole-plot treatments are such that they could not be conveniently applied to such small plots as the sub-plot treatments.

If, however, the whole-plot treatments can be arranged in the form of a Latin square the situation is somewhat different, for if the whole of the treatments were randomized the Latin square arrangement might have to be sacrificed and a usually less efficient randomized block arrangement substituted.

In order to see what differences would result in practice by complete randomization in place of the use of split-plots, the split-plot experiments carried out at Rothamsted and its associated centres were examined. For the randomized block experiments new errors were computed as above. For Latin square experiments the mean of the sums of squares of rows and columns was first added into the whole-plot error, with a corresponding increase in degrees of freedom. Otherwise the procedure was the same as with randomized blocks.

The results are shown in Table V. In the case of randomized blocks there is a considerable gain in accuracy on the whole-plot comparisons, with, as must be the case, a corresponding loss of accuracy on the split-plot comparisons, relatively less in the case of splits into four. In the case of the Latin squares, however, with plots split into two, there is usually a loss of accuracy even on the main-plot comparisons owing to the transition from the Latin square to a randomized block arrangement. In the case of splits into four there is little to choose on the whole-plot comparisons but a definite gain on the split-plot comparisons.

It can, of course, be objected that the treatment of rows or columns as if they were blocks is unfair to the randomized block method, in that more compact blocks would, in fact, be chosen. The results given later in this paper, however, do indicate that Latin square arrangements are in practice markedly more efficient, and it is doubtful if the free choice of blocks would on the average have

TABLE V.

Split-Plot Arrangements.

Percentage information that would have been obtained with ordinary randomized block arrangements.

	Latin Squares.				Randomized Blocks.			
Plots split into:	2.		4.		2.		4.	
Comparisons involving:	Whole Plots.	Split Plots.	Whole Plots.	Split Plots.	Whole Plots.	Split Plots.	Whole Plots.	Split Plots.
Percentage.	Numbers of Experiments.							
0– 20	—	2	—	—	—	—	—	—
20– 40	2	5	1	—	—	3	—	—
40– 60	6	5	1	1	—	1	—	—
60– 80	4	3	—	3	—	2	—	2
80–100	4	4	2	4	—	1	—	1
100–120	3	3	—	1	1	—	—	—
120–140	1	—	5	—	1	—	—	—
140–160	1	—	—	—	1	—	—	—
160–180	1	—	—	—	4	—	—	—
180–200	—	—	—	—	—	—	1	—
200–220	—	—	—	—	—	—	1	—
220–240	—	—	—	—	—	—	1	—
Total	22		9		7		3	

resulted in any appreciable reduction in the errors below those calculated in the construction of the above table.

One point which must be borne in mind when considering the relative efficiency of randomized blocks and Latin squares is the number of error degrees of freedom. The six degrees of freedom for error provided by the 4 × 4 Latin square have long been recognized as inadequate, at least by Fisher. Something of the order of twelve error degrees of freedom would appear desirable in all experimental results which may afterwards have to be considered in conjunction with other material, unless the effects under investigation are large in comparison with their experimental errors. This seems to me to be the governing consideration in favour of the 2 × 2 × 2 confounded arrangement to be described later, as opposed to the use of a 4 × 4 Latin square with split plots, which on a comparison of errors only would appear to be decidedly more efficient. Two 4 × 4 Latin squares with split plots, or a 3 × 2 × 2 system of treatments arranged in a 6 × 6 square with split plots, are not open to this objection and are likely to be a very efficient arrangements; unfortunately, however, they require 64 and 72 plots respectively, which is frequently more than is practicable.

We may therefore conclude that the Latin square arrangement with split plots is a very efficient way of introducing an extra set (especially a pair) of treatments into a factorial system which can otherwise be arranged in the form of a Latin square, provided that it furnishes sufficient error degrees of freedom on the main comparisons.

The treatment of whole blocks of an ordinary experiment with different subsidiary treatments is a modification of the split-plot type of arrangement. Thus different blocks of a fertilizer experiment might be sown with different varieties. Here practically no information is obtained on the average effects of the block treatments, but their interactions with the principal treatments are determined with full accuracy, though it should be noted that differential response to the principal treatments in the different blocks will affect these interactions; this objection is discussed in Section 12. Such a procedure undoubtedly gives a wider inductive basis to the results and for this reason is to be recommended, though it is doubtful if there are many cases in which information on the subsidiary treatments is really not required. In the case of such subsidiary treatments as farmyard manure, however, it might be maintained that a one year's experiment is in any case almost worthless in assessing its value, but that a knowledge of the differences in response to fertilizers in its presence and absence is of considerable interest and importance.

It may be noted here that although whole blocks of a randomized block experiment may be given different treatments, the procedure is inadmissible with the rows or columns of a Latin square, unless the interactions of the principal and subsidiary treatments are to be disregarded, owing to the resultant non-orthogonality of the interactions with the columns or rows respectively. The point is dealt with elsewhere.[8] The rows themselves may be split, *e.g.* for varieties, but this, of course, increases the number of plots, unless varietal differences are ignored at harvesting, and thus is not a particularly efficient arrangement.

Another type of arrangement that stands condemned, but for a different reason, namely, because of a biased error, is the "semi-Latin square." In this arrangement there are twice (or more generally k times) as many rows as columns, and every treatment occurs once in each column and once in each pair of (or each k) rows, which are treated as a unit. It is only possible to make unbiased estimates of the appropriate errors when the treatments are grouped so as to give the equivalent of a split-plot arrangement.

7. *Confounding.*

Confounding is a method of reducing the block size by not completely replicating within each block. We have seen that in a factorial experiment the treatment comparisons are divisible into main effects and interactions of varying complexity. High-order interactions are usually non-existent, or at least small in magnitude compared with their experimental errors, and for this reason, if no other, of little practical interest. Certain components of these high-order interactions appearing in the analysis of variance table (though by no means all) are derived from the direct comparisons of equal groups of the various treatment combinations, all combinations being included. If smaller blocks are used, each containing only the combinations belonging to one group, the contrasts between the groups will coincide with block differences. No information, except a trivial amount derived from inter-block comparisons, is therefore available on the corresponding interactions, which are said to be confounded with blocks.

Provided that the other constituents of treatments in the analysis of variance table are orthogonal with the confounded interactions the remaining comparisons will not be formally affected by the confounding, being entirely intra-block. Their accuracy may be expected to be increased owing to reduction of error due to the smaller blocks.

A very simple example is provided by the $2 \times 2 \times 2$ system of treatments. If the experiment is arranged in blocks of four plots, half of which contain the treatments

$$(1),\ np,\ nk,\ pk,$$

and the other half the treatments

$$n,\ p,\ k,\ npk,$$

the second order interaction will be confounded.

It is easy to see that the main effects and first order interactions are unaffected by block differences. In each two positive and two negative treatment combinations occur in each block.

The experiment on peas at Biggleswade which has already been considered will form a useful example. The plan and yields are given in Fig. 3. The treatment totals are as follows :

Sub-blocks (*a*).				Sub-blocks (*b*).			
(1)	*np*	*nk*	*pk*	*n*	*p*	*k*	*npk*
154·3	173·8	164·0	151·5	191·3	163·0	156·0	163·1

The full analysis of variance is given in Table VI.

The components of the treatment sum of squares may be checked by computing the treatment sum of squares from the treatment totals and also the confounded interaction as if confounding did not exist. These calculations give 384·79 and 37·00 respectively, and the

Block total					Block total
216·1	*pk* 49·5	(1) 46·8	*n* 62·0	*k* 45·5	200·5
	np 62·8	*nk* 57·0	*npk* 48·8	*p* 44·2	
229·8	*n* 59·8	*k* 55·5	*np* 52·0	*nk* 49·8	202·1
	npk 58·5	*p* 56·0	(1) 51·5	*pk* 48·8	
243·1	*p* 62·8	*n* 69·5	*nk* 57·2	*pk* 53·2	225·4
	npk 55·8	*k* 55·0	*np* 59·0	(1) 56·0	

← rows →

Area of each plot : 1/70 acre. (31·75 links × 44·7 link rows.)

FIG. 3.

Experiment on Peas.

Plan and yields in lbs.

difference represents the sum of squares due to treatments in the analysis. The significant results of the experiment have already been discussed.

TABLE VI.

Experiment on Peas.

Analysis of Variance.

	D.F.	Sum of Squares.	Mean Square.
Blocks	5	343·30	68·66
N	1	189·28	—
P	1	8·40	—
K	1	95·20	
N × P	1	21·28	—
N × K	1	33·14	—
P × K	1	0·48	—
Error	12	185·28	15·44
Total	23	876·36	—

Instead of always confounding the same set of interactions it is possible to confound different sets in different sets of blocks.

Such a procedure is called *partial confounding*. Some information will then be obtained on all degrees of freedom, the loss of information being spread over several sets of degrees of freedom instead of being confined to one set.

In a $2 \times 2 \times 2$ system, for instance, each first order interaction might be confounded in a quarter of the blocks, and the second order interaction in the remaining quarter. The advisability or otherwise of such a course depends on the relative importance of the first and second order interactions.

A somewhat more complex example is provided by the $3 \times 3 \times 3$ system. Consider first the 3×3 table:

	a_1.	a_2.	a_3.
b_1	x_1	x_2	x_3
b_2	y_1	y_2	y_3
b_3	z_1	z_2	z_3

There are eight degrees of freedom. The two degrees for each main effect are given by the contrasts of the marginal totals. The four degrees of freedom for interactions are given by similar contrasts of diagonal totals, as follows.

	D.F.	Contrasts between
A main effects	2	$x_1 + y_1 + z_1$, $x_2 + y_2 + z_2$, $x_3 + y_3 + z_3$
B main effects	2	$x_1 + x_2 + x_3$, $y_1 + y_2 + y_3$, $z_1 + z_2 + z_3$
Interactions	2	$x_1 + y_2 + z_3$, $x_2 + y_3 + z_1$, $x_3 + y_1 + z_2$
	2	$x_1 + y_3 + z_2$, $x_2 + y_1 + z_3$, $x_3 + y_2 + z_1$

The degrees of freedom in a $3 \times 3 \times 3$ arrangement are capable of similar subdivision. There are eight degrees of freedom for the second order interactions, which can be split into four sets of two. Just as in a 3×3 table the interactions are given by the contrasts of the diagonal totals, so in a $3 \times 3 \times 3$ table the second order interactions are given by the contrasts of totals of diagonal planes in three dimensions. Thus one of the sets (I) is given by the contrast of all the 1's, all the 2's and all the 3's in the following table:

	a_1.			a_2.			a_3.		
	b_1.	b_2.	b_3.	b_1.	b_2.	b_3.	b_1.	b_2.	b_3.
c_1	1	2	3	3	1	2	2	3	1
c_2	3	1	2	2	3	1	1	2	3
c_3	2	3	1	1	2	3	3	1	2

A second (II) is given by interchanging a_2 and a_3 in this table, and the other two (III and IV) are obtained by interchanging c_2 and c_3 in the first two sets. If the 1's, 2's and 3's of the above table are placed in separate blocks the two degrees of freedom for second order interactions represented by this table will be confounded. An example of this type of confounding is given later in the paper.

Since in a 3 × 3 × 3 system the split of the eight degrees of freedom into four sets of two is purely formal, all sets are of equal importance. Partial confounding should therefore be resorted to, different sets being confounded in different blocks. With 108 plots equal information will be obtained on all four sets, which introduces satisfactory simplicity into the statement of the results.

In the above examples the confounded degrees of freedom belong to the set of highest order interaction degrees of freedom in the ordinary subdivision of the treatment degrees of freedom. Such arrangements are only possible in a limited number of symmetrical systems. In a 3 × 2 × 2 system, for instance, the treatment degrees of freedom are normally partitioned into :

		D.F.			D.F.
Main Effects	A	2	Interactions	A × B	2
	B	1		A × C	2
	C	1		B × C	1
				A × B × C	2

Only the main effects and the interaction B × C can be completely confounded. Any other division of a complete replication involves more than one of the above sets of degrees of freedom.

To avoid sacrificing all information on the possibly important B × C interaction the balanced arrangement given in Table VII

TABLE VII.

3 × 2 × 2 *arrangement.*

Replication :	I.						II.						III.					
Block :	I*a*.			I*b*.			II*a*.			II*b*.			III*a*.			III*b*.		
Treatment :	*a.*	*b.*	*c.*	*a.*	*b.*	*c.*	*a.*	*b.*	*c.*	*a.*	*b.*	*c.*	*a.*	*b.*	*c.*	*a.*	*b.*	*c.*
	0	0	1	0	0	0	0	0	0	0	0	1	0	0	0	0	0	1
	0	1	0	0	1	1	0	1	1	0	1	0	0	1	1	0	1	0
	1	0	0	1	0	1	1	0	1	1	0	0	1	0	0	1	0	1
	1	1	1	1	1	0	1	1	0	1	1	1	1	1	1	1	1	0
	2	0	0	2	0	1	2	0	0	2	0	1	2	0	1	2	0	0
	2	1	1	2	1	0	2	1	1	2	1	0	2	1	0	2	1	1

was devised. In this arrangement B × C is confounded as little as possible in each division. The balance attained by the three replications performs an important function and leads to a considerable simplification of the computations.

In this arrangement as little as $\frac{1}{9}$ of the information on B × C (1 D.F.) is lost, as compared with a similar unconfounded comparison. On the second order interactions A × B × C (2 D.F.)

$\frac{4}{9}$ of the information is lost. It may be noted that $1 \times \frac{1}{9} + 2 \times \frac{4}{9} = 1$, corresponding to the one degree of freedom involved in the division of each replication into two parts. This relation only holds with the balanced arrangement.

Similar balanced arrangements have been devised for the $3 \times 3 \times 2$ system of treatments, and these can be extended to any system of the type 3×2^n and $3 \times 3 \times 2^n$.

Higher degrees of confounding than those exemplified in the above examples are sometimes advantageous, but the degree of confounding that can be indulged in without involving important interactions is very limited. In the 2^5 system, for instance, any single degree of freedom including the fourth order interaction can be confounded by a single division into blocks of 16 plots. If a triple division into blocks of 8 plots is required, at least two second order interactions and one third order of the type $A \times B \times C$, $A \times D \times E$ and $B \times C \times D \times E$ must be confounded; if the fourth order interaction is confounded a first order interaction or main effect is also necessarily involved. While in the case of blocks of four plots at least two first order interactions are necessarily confounded, a possible set being $A \times B$, $C \times D$, $A \times C \times E$, $A \times D \times E$, $B \times C \times E$, $B \times D \times E$, $A \times B \times C \times D$.

A high degree of confounding is therefore not likely to be of any value in agricultural research. In other branches of research, however, where whatever corresponds to block size is more strictly limited it may be of considerable utility. With five replications of a 2^5 system in blocks of four, for instance, a different pair of first order interactions, and their associated second and third order interactions, could be confounded in each replication. The loss of information would then be $\frac{1}{5}$ on all first order, $\frac{2}{5}$ on all second order, and $\frac{1}{5}$ on all third order interactions.

A variant on the standard systems which is of frequent occurrence, and which modifies the strictly formal analysis, both of ordinary factorial experiments and confounded experiments, is the occurrence of dummy treatments. If, for example, we are investigating three levels and three forms of nitrogenous manuring in all combinations, the three plots of each replication receiving no nitrogen are, in fact, identical. In confounded experiments this leads to the occurrence of plots receiving the same treatments in different blocks of the same replication, and these plots can, if desired, be used to furnish information on the differences between these blocks and therefore on apparently confounded degrees of freedom.

It should be mentioned here that although information is available on partially confounded degrees of freedom, its presentation in the form of a table of yields of individual treatment combinations necessitates a certain amount of extra computation. The difficulty does not arise in experiments with certain degrees of freedom totally confounded, nor in partially confounded experiments when we are content to present only the unconfounded effects.

8. *The Estimation of Error from High Order Interactions.*

Another difficulty arising from the large number of treatment combinations in an elaborate factorial experiment is that even two replications of each treatment combination may give a larger number of plots than are required to give the necessary accuracy on the main effects.

To meet this difficulty the device has been introduced of estimating the error from certain unconfounded high order interactions, which from previous experience can be confidently expected to be small in relation to experimental error.

A good example of this procedure is provided by the $3 \times 3 \times 3$ type of arrangement. This is a very suitable arrangement for investigating the responses to the three standard fertilizers, since evidence is obtained on the curvature of the response curves as well as their gradient, knowledge of the former being vital when decisions as to the optional dressing have to be made. A single complete replication demands 27 plots, which are as many as can be conveniently undertaken by most non-experimental farms which are interested in fertilizer trials. Were a minimum of 54 plots to be demanded many possible experiments would be ruled out.

(211) 2575	(121) 2599	(202) 2189
(120) 2472	(220) 2517	(020) 2093
(200) 2517	(022) 2411	(210) 2354
(002) 2403	(110) 2252	(111) 2268
(010) 2220	(212) 2381	(001) 1926
(021) 2252	(201) 2067	(122) 2152
(101) 2295	(102) 2021	(221) 2349
(112) 2362	(011) 1953	(012) 2025
(222) 2434	(000) 1989	(100) 2106

← rows →

Treatments: (211) indicates n_2, p_1, k_1, etc.
Area of each plot: 1/10 acre. (50 links × 200 link rows.)

FIG. 4.

Sugar Beet Experiment at Colwick.
Plan and yields of roots in lbs.

As an example of this type of experiment we may take the experiment on sugar beet at Colwick in 1934. The plan and yields of roots are shown in Fig. 4. The experiment is arranged in blocks of nine plots, so that one of the pairs of degrees of freedom for the second order interaction is allotted to blocks. The error is estimated from the remaining second order interactions and from all the first order interactions except the interactions of the regressions. From physical considerations and practical experience these may be expected to be small in relation to error in an experiment of ordinary accuracy. The partially confounded N regr. × P regr. × K regr. can be separated from error if desired, but this is unlikely to be of importance unless the effects of fertilizers are very marked.

The full analysis of variance is given in Table VIII. N regr. is computed from the square of $S(n_2) - S(n_0)$ divided by 18, N dev. by $S(n_2) - 2S(n_1) + S(n_0)$ divided by 54, and the first order interactions from the functions indicated by the tables :

N regr. × P regr.

	p_0.	p_1.	p_2.
n_0	+1	0	−1
n_1	0	0	0
n_2	−1	0	+1

Divisor : 12

N regr. × P dev.

	p_0.	p_1.	p_2.
n_0	−1	+2	−1
n_1	0	0	0
n_2	+1	−2	+1

36

N dev. × P regr.

	p_0.	p_1.	p_2.
n_0	−1	0	+1
n_1	+2	0	−2
n_2	−1	0	+1

36

N dev. × P dev.

	p_0.	p_1.	p_2.
n_0	+1	−2	+1
n_1	−2	+4	−2
n_2	+1	−2	+1

108

The construction of these tables is obvious if we remember that the linear response to n for p_0 and the mean of all k is given by $S(n_2p_0) - S(n_0p_0)$, with similar expressions for the linear response to n for p_1 and p_2. The linear regression of these three quantities is − 1 times the first, 0 times the second, and + 1 times the third, giving the first table; while the curvature is + 1 times the first and third, and − 2 times the second, giving the second table. The third and fourth tables are derived in a similar manner.

The numerical totals over all levels of k are :

	p_0.	p_1.	p_2.	Total.
n_0	6,318	6,198	6,756	19,272
n_1	6,422	6,882	7,223	20,527
n_2	6,773	7,310	7,300	21,383
Total	19,513	20,390	21,279	61,182

from which the n and p effects and their interactions can be calculated.

The division of the second order interactions has already been described.

TABLE VIII.

Sugar Beet Experiment at Colwick.

Analysis of Variance.

	D.F.	Sum of Squares.	Mean Square.
Blocks (I)	2	244,526	122,263
N regr.	1	247,573	—
N dev.	1	2,948	—
P regr.	1	173,264	—
P dev.	1	3	—
K regr.	1	1,120	—
K dev.	1	2,017	—
N regr. × P regr.	1	660	—
N regr. × K regr.	1	70,687	—
P regr. × K regr.	1	616	—
Error	15	262,298	17,487
Total	26	1,005,712	—

Components of Error.

	D.F.	Sum of Squares.		D.F.	Sum of Squares.
N regr × P dev.	1	41,684	P regr. × K dev.	1	26,136
N dev. × P regr.	1	11,271	P dev. × K regr.	1	28
N dev. × P dev.	1	1,261	P dev. × K dev.	1	972
N regr. × K dev.	1	6,110	N × P × K II	2	388
N dev. × K regr.	1	15,335	N × P × K III	2	5,742
N dev. × K dev.	1	95,230	N × P × K IV	2	58,141

In this experiment the response to both n and p reach the 1 per cent. level of significance, but in neither case is there any evidence of falling off in response with higher dressing nor is there any interaction between the two fertilizers. k shows no significant effects.

9. *Independent Factors.*

The estimation of error from high-order interactions leads naturally to the consideration of the case where all interactions can be assumed to be non-existent. A very simple example of this is provided by the operation of weighing.

Suppose that the weights of seven very light objects require to be determined, and that the apparatus being used necessitates an additional observation for zero correction. Suppose, further, that systematic errors are non-existent, or at least small compared with random errors. The obvious procedure would be to weigh each object separately, and to make an eighth weighing with no object to determine the zero correction. The efficiency, however, can be quadrupled by weighing the objects in groups according to the following scheme :

Weighing No.	Objects weighed.
1	$a + b + c + d + e + f + g$
2	$a + b + d$
3	$a + c + e$
4	$a + f + g$
5	$b + c + f$
6	$b + e + g$
7	$c + d + g$
8	$d + e + f$

In this scheme it will be noticed that every object is weighed four times, and that in the four weighings of a given object every other object is included twice, the remaining four weighings also including every other object twice. Thus object a is included in weighings 1, 2, 3, and 4, which together contain b, c, d, e, f, g twice each. Weighings 5, 6, 7, and 8 contain objects b, c, d, e, f, g twice each, but not a. The difference between the mean of 1, 2, 3 and 4, and the mean of 5, 6, 7 and 8, therefore provides an estimate of the weight of a, and since it is the difference of two means of four weighings the estimate has four times the precision (one-half the standard error) of that given by the ordinary procedure.

The formal analogy of the above scheme and the $2 \times 2 \times 2$ factorial system may be drawn here. If the weighings 1 to 8 are replaced by the treatment combinations *npk*, *np*, *nk*, *n*, *pk*, *p*, *k* and (1) respectively, it will be found that the estimate of the weight of a is transformed into the estimate of the main effect N. Similarly, the estimates of the weights of b, c, d, e, f and g are transformed respectively into the estimates of the main effects P and K, and the interactions $N \times P$, $N \times K$, $P \times K$ and $N \times P \times K$.

Cases in which the interactions are certainly negligible are, in fact, rather rare. Even in the example given above, although weights are undoubtedly additive, systematic errors of the apparatus are likely to complicate the issue. But such experimental systems may be useful in certain preliminary surveys, where there is good reason to believe that interactions are small and where it is desired to include as many factors as possible (some of them perhaps unlikely to produce any effect whatsoever).

The experiments described by Tippett [6] and already referred to in the introduction are of this type. He there considers designs for experiments containing 3, 4 or 5 factors, with five values for each factor, the designs being based on 5×5 Latin, Græco-Latin, and hyper-Græco Latin squares. Not only are the error degrees of freedom composed of interactions of the various factors, but the mean values representing the main effects are not in reality pure main effects, but also contain certain interaction components.

10. *Loss of Efficiency due to Increase in Block Size.*

As far as I know no very extensive investigation has been made of the increase in experimental error due to increased size of block, but Wishart [7] in a paper presented to the Empire Cotton-Growing Corporation Conference in 1934 advised against complex experiments on these grounds. He there concludes :

> " I know that there has been a tendency to proceed to complex experiments with two or more sets of interacting treatments, owing to the flexibility of the arrangement. The danger of loss of efficiency on this ground [the large size of the blocks], and on the ground of inadequate replication, should be guarded against by every experimenter."

The question of inadequate replication has already been dealt with. It has been shown that far from there being any loss of efficiency in factorial designs there is a very considerable gain. It remains to consider how far increase in plot error due to increase in block size is likely to outweigh the very real advantages of factorial design.

The question might be best approached by considering the efficiency of various types of arrangement, and in particular of varying block size, over a series of uniformity trials. This, however, has not been possible up to the present owing to the large amount of work involved.

Some indication of the relative merits of various types of arrangement can be gained from the examination of the results of experiments already carried out. Very little extra work is entailed, since the analyses are already in existence.

A comparison of the efficiency of split plots with that of complete randomization within blocks has already been described in Section 6. Two further comparisons have been made on Rothamsted material, one to determine the loss of efficiency that would result if experiments actually arranged in Latin squares or randomized blocks had been completely randomized, and the second to determine the loss of efficiency that would result if experiments actually confounded had not been confounded.

To investigate the effects of complete randomization it is necessary to make an estimate of what the error mean square would have been had there been no restrictions.

Suppose that the mean squares and degrees of freedom in the analyses of variance are as follows :

Randomized Blocks.			Latin Square.		
	D.F.	Mean Square.		D.F.	Mean Square.
Blocks ...	$q-1$	B	Rows ...	$p-1$	R
Treatments ...	$p-1$	T	Columns ...	$p-1$	C
Error ...	$(p-1)(q-1)$	E	Treatments...	$p-1$	T
			Error ...	$(p-1)(p-2)$	E

If the treatments had been dummy the treatment mean square would equal the error mean square, except for errors of estimation. Replacing T by E the total sum of squares in the case of the randomized block experiment would become

$$(q-1)B + q(p-1)E$$

with $pq-1$ degree of freedom. If there were complete randomization the error mean square with dummy treatments would be derived directly from this. The relative efficiency is therefore given by the ratio

$$\frac{(pq-1)E}{(q-1)B+q(p-1)E}.$$

In the case of the Latin square similar reasoning gives the ratio

$$\frac{(p^2-1)E}{(p-1)R+(p-1)C+(p-1)^2E}=\frac{(p+1)E}{R+C+(p-1)E}.$$

The distribution of the percentage efficiencies for all experiments carried out in 1932 and 1933 is given in Table IX. The mean percentage efficiencies for the two years were as follows. (The numbers in brackets indicate the numbers of experiments.)

	1932.	1933.
Randomized Blocks	72·3 (22)	75·2 (22)
Latin Squares	54·1 (38)	57·4 (37)

The differences between the randomized block and Latin square arrangements are quite marked, especially with the larger sizes of blocks. Whereas the randomized block arrangements have only removed on the average 26 per cent. of whatever variation existed over the experimental site, Latin squares have accounted for no less than 44 per cent.

On the basis of these figures, therefore, one might expect a randomized block arrangement to have on an average an efficiency of only $\frac{56}{74}$, or about 75 per cent. of a Latin square on the same plots.

TABLE IX.

Percentage Efficiency with Complete Randomization.

	Latin Squares.			Randomized Blocks.		
	4 × 4.	5 × 5.	6 × 6 and 8 × 8.	4–6 treatments.	7–9 treatments.	10–24 treatments.
Percentage Efficiency.	Number of Experiments.					
0– 10	—	—	—	1	—	—
10– 20	5	2	—	—	—	—
20– 30	3	1	1	2	—	—
30– 40	8	7	—	1	—	1
40– 50	6	3	1	2	—	—
50– 60	4	2	1	2	1	—
60– 70	2	2	—	1	6	1
70– 80	5	4	—	4	1	1
80– 90	7	2	1	—	3	3
90–100	2	1	—	2	4	1
100–110	2	—	1	2	3	2
110–120	2	—	—	—	—	—
Total	46	24	5	17	18	9

A comparison of the average percentage standard errors given in Table X reveals an even greater advantage in favour of the Latin square. (Rothamsted and Woburn experiments have been excluded from this table, since in the later years they consisted of large factorial experiments not comparable with those of the outside

TABLE X.

Average Percentage Standard Errors per Plot.

Outside Centres.

	1927–30.	1931.	1932.	1933.	Weighted Mean.
Potatoes :—					
Randomized Blocks ...	9·0 (6)	10·2 (6)	4·5 (3)	10·6 (7)	9·2 (22)
Latin Squares	5·2 (15)	6·6 (11)	7·8 (15)	7·4 (15)	6·8 (56)
Sugar Beet, Roots :—					
Randomized Blocks ...	7·4 (5)	5·8 (2)	8·2 (7)	12·9 (1)	7·9 (15)
Latin Squares	6·4 (8)	5·3 (5)	7·1 (7)	5·5 (8)	6·1 (28)

centres.) In the case of potatoes the relative efficiency as judged by the ratio of the squares of the mean errors is 55 per cent., and with sugar beet it is 60 per cent. It is to be expected that the apparent efficiency of randomized blocks as compared with Latin squares would be found to be lower when judged on this comparison, as many of the Latin squares were only 4 × 4, so that presumably there was less fertility variation to remove than in the case of the larger randomized block experiments.

On the basis of these figures we can make a rough estimate of the relative efficiency of a factorial design and the equivalent simple arrangements. As an example we will consider an experiment on three levels of each of the three standard fertilizers. With 54 plots we should have the option of making three separate experiments of 18 plots each, or of carrying out a $3 \times 3 \times 3$ factorial experiment in six blocks of 9 plots (two replicates). The separate experiments might well be laid out in two 3×3 Latin squares each (six squares in all) with a pooled estimate of error (eighteen degrees of freedom) from the three experiments, provided these were all in the same field. Taking the error variance per plot in the Latin square arrangement as 50 per cent. of that in the factorial arrangement (*i.e.* somewhat less than the values of 55 per cent. and 60 per cent. given in the last paragraph), the relative efficiency on the main effects will be in the ratio of 3 : 2 in favour of the factorial design (since three times as many plots are involved in all means). Thus in mere accuracy of estimation of main effects the factorial design is 50 per cent. more efficient, as well as having the additional and even more important advantage of providing information on the interactions and a wider inductive basis for the results.

It will be noted that the factorial arrangement might be expected to be even more efficient relatively to the simple arrangements if these latter had been arranged in randomized blocks instead of Latin squares. It is not so much the increase in size of block in a complex experiment that causes a higher error variance per plot as the necessity of abandoning the admittedly more efficient Latin square arrangement. Indeed, as we have seen, that complete randomization without restrictions of the 1932 and 1933 randomized block experiments would only have decreased the efficiency by about 25 per cent. on the average.

The examination of the confounded experiments confirms this. The reduction in block size due to the confounding has resulted on the average in quite moderate increases in efficiency.

In making the estimation of loss of efficiency due to failure to confound it must be assumed that had the experiment not been confounded the blocks composing each complete replication would have formed one large block; it is, of course, possible that the experimenter might have laid out the experiment with differently shaped large blocks had he not been confounding. To simplify the calculations it has also been assumed that partially confounded interactions are negligible except where they have been judged significant. Completely confounded interactions have always been assumed negligible.

The method of calculation is as follows. Suppose that each block contains k plots, and that there are hk treatment combinations in all, so that in each replication $h - 1$ degrees of freedom are confounded. Let there be r replicates. The analysis of variance with a single set of degrees of freedom completely confounded is as follows.

	D.F.	Mean Square.
Blocks { Between complete replications... ...	$r - 1$	A
Blocks { Within ,, ,, ...	$r(h - 1)$	B
Treatments...	$h(k - 1)$	T
Error	$h(r - 1)(k - 1)$	E

Replacing T by E and including the second part of blocks gives the sum of squares

$$r(h - 1)B + rh(k - 1)E,$$

with $r(hk - 1)$ degrees of freedom, so that the percentage efficiency without confounding is

$$\frac{(hk - 1)E}{(h - 1)B + h(k - 1)E}.$$

Table XI gives the distribution of this fraction for various groups of Rothamsted experiments. In less than 10 per cent. of

TABLE XI.

Relative Efficiency without Confounding.

	2 × 2 × 2.		3 × 3 × 3.		Miscel-laneous.	Total.
	Small Plots.	1/10 acre Plots.	Small Plots.	1/10 and 1/20 acre Plots.		
Percentage Efficiency.	Number of Experiments.					
0– 10	—	—	—	—	—	—
10– 20	—	—	—	—	—	—
20– 30	—	1	—	—	—	1
30– 40	1	—	1	—	—	2
40– 50	—	1	1	—	—	2
50– 60	—	—	3	—	2	5
60– 70	—	4	1	1	2	8
70– 80	4	1	5	2	2	14
80– 90	2	3	1	—	3	9
90–100	2	1	7	2	1	13
100–110	2	—	2	4	—	8
110–120	2	2	—	—	—	4
Total	13	13	21	9	10	66

the experiments has the efficiency been more than doubled by the confounding, but in over half the experiments it has been raised by more than 25 per cent. In about a third of the experiments the gain was either trivial or non-existent.

11. *A Further Consideration Governing Optimal Block Size.*

There is another aspect of the question of the most efficient block size which is of importance. It is often tacitly assumed that the block size which gives the minimal error variance per plot will be the most efficient, but in fact it is necessary to take into consideration not only the variation of the error with change in block size, but also the number of treatments it is desired to test.

The points involved can be brought into prominence by considering the relative efficiency of the (randomized) half-drill strip method and the randomized block or Latin square method of comparing several treatments. The discussion applies equally to any system of comparing the treatments in pairs.

Usually in the method of pairs one treatment is chosen as control and all others are compared with it. This has one obvious disadvantage, in that the comparisons with control are made with twice the accuracy (*i.e.* $1/\sqrt{2}$ of the standard error) of comparisons between any other pair of treatments, since if A is the control, the difference of B and C must be deduced from $\delta_m(B - A) - \delta_m(C - A)$, where $\delta_m(B - A)$ is the mean difference of B and A in the pairs where they occur together. This disadvantage can be obviated by making comparisons between every pair of treatments. Thus with five treatments, A, B, C, D, E, there are ten possible comparisons AB, AC, AD, AE, BC, BD, BE, CD, CE, DE. It is a remarkable fact that with this type of design, using the same number of plots, the same accuracy is obtained on all comparisons as would be obtained on the comparisons of the control and another treatment in the customary type of arrangement.

With the symmetrical arrangement, if σ'^2 is the error variance per plot, p the number of treatments, and n the available number of plots, the variance of the difference between two treatment means is

$$\frac{4(p-1)}{n}\sigma'^2.$$

With the " control " arrangement this is the variance of the $(p - 1)$ comparisons involving the control, while the other $\frac{1}{2}(p - 1)(p - 2)$ comparisons have double the variance. The mean variance of all comparisons is therefore

$$\frac{8(p-1)^2}{pn}\sigma'^2.$$

With an arrangement in randomized blocks or Latin squares, with an error variance of σ^2 per plot, the variance between two treatment means is

$$\frac{2p}{n}\sigma^2.$$

Thus, provided that σ^2/σ'^2 is less than $2(p-1)/p$, the randomized block or Latin square arrangement is more efficient than the symmetrical arrangement, while σ^2/σ'^2 must be greater than $4(p-1)^2/p^2$ for the control method to be the more efficient. With six treatments these fractions have the values of 5/3 and 25/9 respectively.

In agricultural field trials the greater similarity of pairs of plots over those of blocks of five is seldom likely to give a ratio of σ^2/σ'^2 as great as 25/9, or even 5/3.

Tables XII and XIII show the results of a test on two uniformity trials. In Mercer and Hall's wheat data [5] randomized blocks gave

TABLE XII.

Mercer and Hall. Wheat.

Plots used: 5 unit-plots along rows (E and W) × 1 unit-plot across rows, giving 100 plots (5 × 20) of 1/100 acre each.
Blocks of 5 plots across rows, or four 5 × 5 Latin squares.

	Error mean square.
Randomized blocks	0·940
Latin squares	0·723
Half-drill strip method	0·698
Randomized pairs	0·742

TABLE XIII.

Immer. Sugar Beet.

Plots used: 4 rods long × 4 rows wide, 2 edge rows rejected.
Rows 7–54 only used: 60 plots (5 × 12) of 1/90 acre each.
Blocks of 6 plots across rows, or two incomplete 6 × 6 Latin squares.

	Error mean square.
Randomized blocks	237·2
Latin squares	114·4
Half-drill strip method	130·6
Randomized pairs	161·8

an error mean square of 0·940, compared with 0·698 by the half-drill strip method, their ratio being 1·35. The method of randomized blocks is therefore much more efficient.

In Immer's sugar beet data [4] randomized blocks gave 237·2 and the half-drill strip method 130·6 with a ratio of 1·82. This is slightly greater than 5/3, but much less than 25/9. The symmetrical method would therefore be slightly more efficient than randomized blocks, but the control method considerably less so.

In both cases the Latin square arrangements have reduced the error to that of the half-drill strip method, so that the latter method must be judged very inferior to the use of Latin squares. In one case only has the half-drill strip method any appreciable advantage over random pairs. In fairness, however, it should be pointed out that the plots are wider than they would properly be on the half-drill strip method.

In other experimental material much greater similarity may exist between pairs than between larger blocks, as, for example, in human beings, where the similarity between monozygotic twins is strikingly greater than between any other groups that can be formed. In such cases the arrangement in symmetrical pairs may be of considerably greater efficiency than any arrangement based on the use of larger blocks.

It may be mentioned here that the method of symmetrical pairs is capable of extension to blocks of any number less than the number of treatments. Thus if the experimental material naturally forms blocks of three, six treatments may be tested by an arrangement such that (1) every treatment occurs equally frequently and not more than once in each block, (2) every pair of treatments occur together in the same number of blocks. For this type of arrangement I propose the name of *symmetrical incomplete randomized blocks*. Certain modifications are required in the analysis of variance, and in the presentation of results, but these modifications are surprisingly simple and demand very little extra computational labour.

12. *Differential Responses in Different Blocks.*

An objection that has sometimes been made against confounding is that the effects of one or more of the treatments may vary from block to block. This will tend to inflate some other interaction sum of squares. In the $2 \times 2 \times 2$ arrangement, for instance, if blocks containing the *npk* plots respond better to nitrogen than those containing no fertilizer it is easy to see that an apparent positive $P \times K$ interaction will result. There will, of course, be no bias in this interaction, for positive and negative values will occur with equal frequency, but it will be judged significant more often than it should if compared with the ordinary error.

The objection is not really one against confounding, but against the whole system of pooling estimates of error in the analysis of variance. If such variation in response exists, then the error is no longer homogeneous, and should be split up into its component parts, namely, the interaction of each treatment degree of freedom with blocks. The response to *n*, for instance, would then be tested against the variation of this response from block to block, namely

$N \times$ blocks, and in so far as the blocks could be regarded as a random sample of all possible blocks in the field, this would answer the question as to whether the response to nitrogen over the field could be regarded as significant or whether it was due to a lucky selection of blocks. This subdivision of errors cannot be made in Latin square arrangements.

In practice the force of the objection depends entirely on whether, in fact, there is likely to be any substantial variation in response from block to block in experiments ordinarily carried out. It is well known that response to fertilizers varies very greatly from farm to farm, and it would therefore appear possible that such variations occur even on different parts of the same field.

There are various methods of testing whether, in fact, this is so. One is to tabulate the mean squares of the interactions of blocks with each treatment degree of freedom, and see if they are homogeneous by testing against χ^2. Again, in confounded experiments the interaction of some particular main effect, say N, with sub-blocks can be compared with the remaining error degrees of freedom. Yet a third way, when there are split plot experiments available, is to examine the interaction of the split treatment with blocks or with the rows and columns of a Latin square.

This last test has been carried out on the experiments with split plots at Rothamsted and its associated centres. Only those experiments in which the split treatment shows a significant effect were chosen. In each case the probability of getting as large or larger apparent effect was calculated from the z distribution. Table XIV shows the results. In Latin squares rows and columns have been taken together if the error degrees of freedom were less than 40.

TABLE XIV.

Interaction with Soil Differences.

Year.	Place.	Crop.	Treatment.	Interaction with	n_1.	n_2.	P.
1929	Rothamsted (two squares and double	Barley for single nitrogen)	Potash	Rows	4	52	0·884
				Columns	4	52	0·878
				Rows	4	52	0·340
				Columns	4	52	0·752
1931	Rothamsted	Wheat	Harrowing	Blocks	3	24	0·157
,,	,,	Oats	Nitrogen	Blocks	5	40	0·614
,,	Badminton	Hay	Potash	Rows and Cols.	8	12	0·519
,,	Chesterfield	Hay	Potash	Rows and Cols.	8	12	0·415
,,	Downham	Potatoes	Nitrogen	Blocks	3	24	0·424
,,	,,	Sugar Beet	Nitrogen	Blocks	3	24	0·286
,,	Potton	Potatoes	N/S–S/A	Rows and Cols.	6	6	0·254
,,	,,	Cabbages	N/S–S/A	Rows and Cols.	6	6	0·245
1932	Rothamsted	Potatoes	Nitrogen	Blocks	2	94	0·155
,,	Colchester	Sugar Beet	Salt	Blocks	3	24	0·020
,,	Oakerthorpe	Mangolds	Dung	Rows and Cols.	6	6	0·349
1933	Rothamsted	Wheat	Nitrogen	Rows	3	41	0·530
				Columns	3	41	0·640
,,	,,	Forage	Nitrogen	Rows and Cols.	8	12	0·116
,,	Elsham	Sugar Beet	Nitrogen	Rows and Cols.	8	12	0·896

The table shows no evidence of any variation in response. There is only one significant result ($P = 0{\cdot}0195$), that of salt in the 1932 experiment at Colchester, and examination of the yields of the individual plots of this experiment did not support the theory of differential response.

The distribution of the probabilities is shown in Fig. 5. The

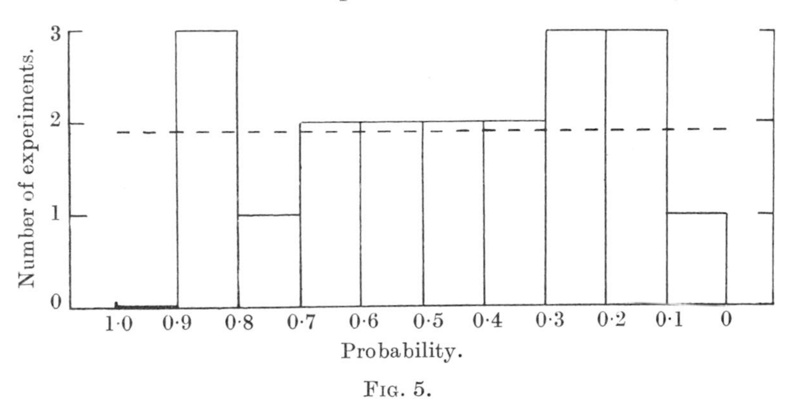

FIG. 5.

Interactions of treatments with soil differences: distribution of probabilities.

The dotted line shows the expectation in each class.

probability of obtaining a set of probabilities lower than these by chance is given by the sum of half the Napierian logarithms of the probabilities, which will be distributed as χ^2 with 38 degrees of freedom.[1] This probability was found to be 0·34.

13. *Conclusions.*

I hope that the results presented in this paper will be a sufficient demonstration that, rightly used, the complex methods of experimentation included in the term factorial design are of considerable value in agronomic research. The special devices of confounding and the estimation of error from high-order interactions admittedly complicate the statistical analysis of the results, but many of the difficulties vanish with familiarity, and others (such as the somewhat laborious calculations required to furnish a table of yields of all treatment combinations freed from block effects) arise from the attempt to present the results in a form which is excellent for an ordinary randomized block or Latin square experiment but unsuitable for a confounded arrangement.

There remains the danger of misinterpretation owing to faulty statistical analysis, and the perhaps more serious danger of faulty design. Both are real dangers, but the adoption of standard arrangements would appear to be an adequate safeguard. In practice most workers are unable to give special attention to the design and analysis of each individual experiment, and if only for this reason the use of standard patterns appears to be a necessity.

I have not attempted to discuss the relative practical difficulties of factorial and simple experiments. Many of these difficulties undoubtedly arise from unfamiliarity and vanish completely as soon as one or two factorial experiments are actually carried out. Of those that do not, the man actually responsible for the conduct of the experiments in the field must be the judge. At the same time, in order to judge fairly he must be able to assess the advantages of factorial design.

Factorial experiments, particularly confounded experiments, are more vulnerable than simple experiments with small blocks. The failure of any considerable part of a single block involves the results of the whole of that replication in which it occurs, and any use of the replication necessitates tiresome and lengthy computations. The failure of animals in animal husbandry experiments can be equally troublesome. For this reason excessively complex factorial experiments should not be undertaken when the experiments are exposed to serious natural hazards.

An important advantage of the straightforward randomized block arrangement is that an estimate of error can be isolated for every comparison separately. This is of very great value when handling new and unknown material, or treatments which may produce large differences and even partial or complete failures. In such cases the assumption of constancy of error variance is entirely unjustified, but in a randomized block experiment any treatment or treatments may be excluded and the analysis carried out on the remainder. This is not true of either the Latin square or of confounded arrangements.

I have already expressed the opinion that factorial design is likely to be of interest to workers in other fields of research, and that in many of these fields the complication of confounding is likely to be unnecessary. At Rothamsted we are using factorial designs in experiments with pigs, and also in pot culture work. I have not space to say more of these applications here, but I would like to conclude by expressing the hope that workers in other fields may find these methods as effective as they have already proved in agricultural research.

References.

[1] Fisher, R. A. *Statistical Methods for Research Workers* (1925). Edinburgh: Oliver and Boyd (5th Edition, 1935).

[2] Fisher, R. A. The arrangement of field experiments. *Jour. Min. Agric.* (1926), vol. 33, pp. 503–513.

[3] Fisher, R. A., and Wishart, J. The arrangement of field experiments and the statistical reduction of the results. Imperial Bureau of Soil Science, Technical Communication (1930), No. 10.

[4] Immer, F. R. Size and shape of plot in relation to field experiments with sugar beets. *Jour. Agric. Res.* (1932), vol. 44, pp. 649–668.

[5] Mercer, W. B., and Hall, A. D. The experimental error of field trials. *Jour. Agric. Sci.* (1911–12), vol. 4, pp. 107–132.

[6] Tippett, L. H. C. Some applications of statistical methods to the study of variation of quality of cotton yarn. *Jour. R.S.S. Supplement* (1935), vol. 2, pp. 27–62.

[7] Wishart, J. Analysis of variance and analysis of co-variance, their meaning, and their application in crop experimentation. Empire Cotton-Growing Corporation Conference, July 1934. Report, pp. 83–96.

[8] Yates, F. The principles of orthogonality and confounding in replicated experiments. *Jour. Agric. Sci.* (1933), vol. 23, pp. 108–145.

AUTHOR'S WRITTEN REPLY TO DISCUSSION

Mr. Yates added the following written reply:

I am somewhat alarmed, on reading the remarks of the speakers at the meeting, and of others who did not speak, to note the considerable accretions that have occurred since the actual discussion. I must plead forgiveness if my reply should appear, by comparison, somewhat brief and terse.

A general survey of the remarks immediately brings out one fact of considerable interest. Those speakers who are actually engaged in experimental work or are in close contact with experimental workers, appear completely satisfied with factorial design, while those who are not so engaged have raised several objections to the method. Now if the method of factorial design were as fundamentally unsound, misleading, and unreliable as the critics would have us believe, one would have expected some at least of those who had most used it to have discovered the fact, both from bitter experience and because they had at one time and another devoted a considerable amount of thought to it. But this is not the case. I am encouraged, therefore, to believe that the force of these criticisms is not perhaps as great as their volume would indicate.

I will now deal with the remarks of the various speakers in detail. Mr. Bartlett's confounding of quantities of nitrogen, of which he gives a detailed plan, is a good example of the possibilities of this type of arrangement. It is similar to many Rothamsted experiments involving quality factors, though we have not usually taken the confounding so far. I also like his suggestion for a combined variety and $2 \times 2 \times 2$ *npk* experiment.

I have myself frequently considered a $2 \times 2 \times 2$ lay-out in two 4×4 Latin squares, but, like Mr. Bartlett, have so far been deterred from using this lay-out because of the possibility of differential responses.

With regard to the question raised by Mr. Bartlett of what are the most appropriate interactions to use for an estimate of error in experiments with only a single replication, it would certainly appear that if we know that the first response $A = S(n_1 - n_0)$ to a factor n is going to be large in comparison with the second $B = S(n_2 - n_1)$, it would be best to include in error the interactions of B with the other factors and to set out for examination the interactions of $\frac{1}{2}(n_1 + n_2) - n_0$, which are orthogonal to the other set. It would also appear that with normally distributed errors it would be quite legitimate to take into account the experimental results on the main effects when deciding what interactions should be set out for examination and what should be included in error. It is not, however, clear to me that this process will be legitimate if there are one or two outstanding plots which differ widely from the rest. The use of this criterion, therefore, requires caution, though I have found it satisfactory on one occasion, where there was a striking response to a single dressing of nitrogen, with marked interactions, but no further response to double and treble dressings. In this case the legitimacy of the criterion was confirmed by other results of the same experiment for which there was proper replication.

Mr. Tippett's arrangement in an incomplete Latin square has, in fact, nothing in common with a semi-Latin square. The estimate of error is unbiased (in the sense developed later) in the case when only *one* row is missing, and in this case a neat least-square solution exists; the same is true when one row and one column, or either and one treatment, are missing. Thus such designs are both valid and useful. (Dr. Crowther has drawn attention to their existence in his concluding remarks.) If more than one row is missing, however, the least-square solution is unmanageable, and the investigation of bias is therefore of little practical interest.

In saying that in Mr. Tippett's arrangements the mean values representing the main effects contained certain interaction components I was using interaction to imply interactions between factors in which we were interested. There is a difference between such interactions and those with factors we wish to ignore, such as soil effects. Nor in general is the error term in an agricultural experiment anything more than error, since true interactions of soil and treatment are very small within the limits of a single experiment, as I have shown in Section 12. Therefore I think that "error" is a satisfactory term.

Dr. Crowther's point about the results of split-plots experiments being on the face of them misleading is an important one which I had forgotten when assessing the relative merits of splitting and confounding. In the summary tables of a split-plot experiment a factor a may appear to produce much the same effect in the presence of all levels, 1, 2, 3, 4 of another factor b for which the plots are split without the average effect of a being significant.

I am sorry if Dr. Irwin found the paper in part unintelligible. This does not seem to have been the experience of other workers. I suspect that Dr. Irwin did not follow the advice given on p. 183, and that had he done so most of his difficulties would have vanished.

I am grateful to Professor Fisher for setting out fully the two-way distribution of degrees of freedom, corresponding to the topographical and experimental divisions of a Latin square. The origin of the bias of error in the semi-Latin square is easily seen from this table. If the mean square error between plots is E and that within plots is E', and the treatments produce no effect, then in a 5×5 semi-Latin square which is of the split-plot type the total treatment sum of squares (9 degrees of freedom) will have an expectation of

$$4\,E + 5\,E',$$

so that the expectation for the general treatment mean square is

$$\tfrac{4}{9}\,E + \tfrac{5}{9}\,E'.$$

Similarly, the pooled estimate of error will be

$$\tfrac{12}{32}\,E + \tfrac{20}{32}\,E',$$

giving a bias for a 5×5 square of

$$\tfrac{5}{72}\,(E - E').$$

In a semi-Latin square which is not of the split-plot type this pooling cannot in practice be avoided, and it appears on examination that the bias will be of the same magnitude as that just given.

It is interesting to consider the result of applying a similar line of argument to a randomized block arrangement with split plots. Here it will be found that no bias of the above nature exists, as should clearly be the case, since the final arrangement is equivalent to an arrangement in simple randomized blocks.

I have already answered most of Dr. Wishart's points in my reply at the meeting, but in passing I would emphasize that it is not a logical conclusion of my results that we can make the blocks as large as we please, nor could I defend factorial design at the cost of condemning the randomized block method, since such a condemnation would automatically condemn factorial design, which must ordinarily use randomized blocks. I should, of course, be the first to be pleased if it were possible to make randomized blocks as accurate as Latin squares, but for reasons stated at the time I doubt if Dr. Neyman has shown us the way.

Both Dr. Wishart and Dr. Neyman have raised the objection that an interaction may exist and be large enough to make the effects of a given factor in the presence and absence of another factor decidedly different from the main effect, without the interaction itself being significant. I have discussed just such a situation in Section 4, where I conclude by saying :

> in the absence of consistent interactions of this type over a series of experiments and of any knowledge of when they are likely to occur, the average response to n and the average depression with k are the best estimates to adopt in assessing the advantages of these fertilizers.

Dr. Neyman has given his ideas of what will happen when, as he says, "Nature chooses to be frivolous"; though I think that Dr. Neyman is perhaps over-stressing cases which rarely occur in Nature. What Dr. Neyman has overlooked is my phrase "a series of experiments," and also the phrase "in the absence of any consistent interactions of this type." Indeed the situation constructed by

Dr. Neyman, if it arose over a series of experiments, would puzzle no one, and no false recommendations would be made. Results would be obtained of the type obtained by Dr. Neyman by his sampling process. They would be set out in a table just as he has set out his in Table II, together with the other main effects and interactions. The means of each column would be calculated, and by a statistical process which I need not go into here the verdict would be given as follows :

A definitely significant average main effect of a equal to 1·74 [Dr. Neyman's "true" value 1·755] and definitely significant average interactions $a \times b$ equal to 1·46 ["true" value 1·545], and $a \times c$ equal to 1·53 ["true" value 1·545]. None of these effects is significantly different for the different experiments.

Even, however, if we have only the results of a single factorial experiment on a given set of factors, these results cannot be misleading in the way Dr. Neyman suggests. The main effect of the factor a is the average of the effects of a in the presence and absence of all combinations of the other factors b and c, and it is on this main effect that the experiment furnishes an answer of high precision. The experiment also furnishes information of known, but lower, precision on the separate effects of a in the presence and absence of each combination of b and c (Dr. Neyman's "simple" effects), but it has never been claimed that the simple effects are identical with the main effects. To assume this and to base recommendations on such an assumption implies knowledge other than that provided by the experiment.

In general the range of conditions in which the response to the factor a is investigated should be such as to be included in the range of conditions found in practice, but in certain cases it may be advisable to extend the experimental range in order to investigate questions of scientific interest. In agricultural experimentation there is little danger of overstepping the practical range, since practical conditions are so variable.

In passing it may be noted that Dr. Neyman has exaggerated the importance of his interactions. He has forgotten that they should be divided by a factor 2 when being added to or subtracted from the main effect to give the effect of a factor a in the presence of or absence of b and c. (Since the meeting I have altered my definition of an interaction by the inclusion of a factor $\frac{1}{2}$, for reasons stated in the text.)

Actually in manurial experiments results of the simplicity of those given in Dr. Neyman's table are rare. The main effects, if they are of any magnitude, differ significantly from experiment to experiment. Factors having large main effects do show occasional interactions of sufficient magnitude to be of importance, but such interactions are rarely consistent, and their average value is small.

This variability in response is a sufficient argument against undue replication. It is of little use determining the value of a manurial response with extreme precision in a given field and a given year if we know, as we do, that the response in a neighbouring field the next year will be widely different. It would doubtless be pleasant to know every experimental result with high accuracy, but the Law of Diminishing Returns operates in the experimental field as in economics, and the practical scientist should seek to conduct his research with a standard of efficiency at least not lower than would be acceptable in a similar commercial undertaking.

In general all arguments which attempt to show that factorial design is ineffective or misleading when interactions exist contain their own refutation, since it is only possible to assess interactions by experiments of the factorial type. If a precise knowledge of the whole set of "simple" effects that go to make up a certain main effect is required, then that main effect is inevitably determined with correspondingly greater precision, and at the same time all the interactions are determined with equal precision. Dr. Neyman has listed an imposing questionnaire, presumably so as to create the impression that the energies of the experimenter are being wasted in attempting to answer all these questions, but he has carefully refrained from making any suggestions as to how the experimenter could possibly answer the questions on the "simple effects" (which Dr. Neyman presumes are his sole interest) without at the same time evaluating the main effects and interactions.

PAPER V

THE ANALYSIS OF GROUPS OF EXPERIMENTS

FROM THE JOURNAL OF AGRICULTURAL SCIENCE
VOLUME XXVIII, PART IV, pp. 556–580, 1938

Author's Note

The results of a group of experiments of identical design can be set out in a multiway table, to which the procedure of the analysis of variance similar to that for a single replicated experiment can be applied. Failure to realize the conditions that must be satisfied for such an analysis to be valid—in particular, approximate equality of errors in the different experiments and homogeneity of the treatment degrees of freedom—led to uncritical and uninformative use of this type of analysis. Paper V suggests ways of making a more meaningful analysis of such experiments, particularly varietal trials and multifactorial fertilizer experiments. The separate analysis of single degrees of freedom formed by linear functions of the treatment means of the different experiments is of particular interest.

The device of taking the mean yields of all varieties in an experiment as a measure of fertility and regressing the yields of individual varieties on this, which is also described in the paper, has recently proved of value to plant breeders in studying environmental and genotype-environmental components of variability (Perkins, Jean M., and Jinks, J. L., *Heredity*, **23**, 339-356, 1968).

1969 FRANK YATES

THE ANALYSIS OF GROUPS OF EXPERIMENTS

By F. YATES and W. G. COCHRAN

Statistical Department, Rothamsted Experimental Station, Harpenden

(With One Text-figure)

1. Introduction

Agricultural experiments on the same factor or group of factors are usually carried out at a number of places and repeated over a number of years. There are two reasons for this. First, the effect of most factors (fertilizers, varieties, etc.) varies considerably from place to place and from year to year, owing to differences of soil, agronomic practices, climatic conditions and other variations in environment. Consequently the results obtained at a single place and in a single year, however accurate in themselves, are of limited utility either for the immediate practical end of determining the most profitable variety, level of manuring, etc., or for the more fundamental task of elucidating the underlying scientific laws. Secondly, the execution of any large-scale agricultural research demands an area of land for experiment which is not usually available at a single experimental station, and consequently much experimental work is conducted co-operatively by farmers and agricultural institutions which are not themselves primarily experimental.

The agricultural experimenter is thus frequently confronted with the results of a set of experiments on the same problem, and has the task of analysing and summarizing these, and assigning standard errors to any estimates he may derive. Though at first sight the statistical problem (at least in the simpler cases) appears to be very similar to that of the analysis of a single replicated trial, the situation will usually on investigation be found to be more complex, and the uncritical application of methods appropriate to single experiments may lead to erroneous conclusions. The object of this paper is to give illustrations of the statistical procedure suitable for dealing with material of this type.

2. General considerations

Agronomic experiments are undertaken with two different aims in view, which may roughly be termed the technical and the scientific. Their aim may be regarded as scientific in so far as the elucidation of the underlying laws is attempted, and as technical in so far as empirical rules for the conduct of practical agriculture are sought. The two aims are, of course, not in any sense mutually exclusive, and the results of most well-conducted experiments on technique serve to add to the structure of general scientific law, or at least to indicate places where the existing structure is inadequate, while experiments on questions of a more fundamental type will themselves provide the foundation of further technical advances.

In so far as the object of a set of experiments is technical, the estimation of the average response to a treatment, or the average difference between varieties, is of considerable importance even when this response varies from place to place or from year to year. For unless we both know the causes of this variation and can predict the future incidence of these causes we shall be unable to make allowance for it, and can only base future practice on the average effects. Thus, for example, if the response to a fertilizer on a certain soil type and within a certain district is governed by meteorological events subsequent to its application the question of whether or not it is profitable to apply this fertilizer, and in what amount, must (in the absence of any prediction of future meteorological events) be governed by the average response curve over a sufficiently representative sample of years. In years in which the weather turns out to be unfavourable to the fertilizer a loss will be incurred, but this will be compensated for by years which are especially favourable to the fertilizer.

Any experimental programme which is instituted to assess the value of any particular treatment or practice or to determine the optimal amount of such treatment should therefore be so designed that it is capable of furnishing an accurate and unbiased estimate of the average response to this treatment in the various combinations of circumstances in which the treatment will subsequently be applied. The simplest and indeed the only certain way of ensuring that this condition shall be fulfilled is to choose fields on which the experiments are to be conducted by random selection from all fields which are to be covered by the subsequent recommendations.

The fact that the experimental sites are a random sample of this nature does not preclude different recommendations being made for different categories included in this random sample. We may, for instance, find that the response varies according to the nature of the previous crop, in which case the recommendations may be correspondingly varied. Moreover, in a programme extending over several years, the recommendations may become more specific as more information is accumulated, and the experiments themselves may be used to determine rules for the more effective application of the treatments tested, as in fertilizer trials in which the chemical examination of soil samples may lead to the evolution of practical chemical tests for fertilizer requirements.

At present it is usually impossible to secure a set of sites selected entirely at random. An attempt can be made to see that the sites actually used are a "representative" selection, but averages of the responses from such a collection of sites cannot be accepted with the same certainty as would the averages from a random collection.

On the other hand, comparisons between the responses on different sites are not influenced by lack of randomness in the selection of sites (except in so far as an estimate of the variance of the response is required) and indeed for the purpose of determining the exact or empirical natural laws governing the responses, the deliberate inclusion of sites representing extreme conditions may be of value. Lack of randomness is then only harmful in so far as it results in the omission of sites of certain types and in the consequent arbitrary restriction of the range of conditions. In this respect scientific research is easier than technical research.

3. The analogy between a set of experiments and a single replicated trial

If a number of experiments containing the same varieties (or other treatments) are carried out at different places, we may set out the mean yields of each variety at each place in the form of a two-way table. The marginal means of this table will give the average differences between varieties and between places. The table bears a formal analogy to the two-way table of individual plot yields, arranged by blocks and varieties, of a randomized block experiment, and we can therefore perform an analysis of variance in the ordinary manner, obtaining a partition of the degrees of freedom (in the case of six places and eight varieties, for example) as follows:

	Degrees of freedom
Places	5
Varieties	7
Remainder	35
Total	47

The remainder sum of squares represents that part of the sum of squares which is due to variation (real or apparent) of the varietal differences at the different places. This variation may reasonably be called the *interaction* between varieties and places. It will include a component of variation arising from the experimental errors at the different places.

If the experiments are carried out in randomized blocks (or in any other type of experiment allowing a valid estimate of error) the above analysis may be extended to include a comprehensive analysis of the yields of the individual plots. If there are five replicates at each place, for example, there will be 240 plot yields, and the partition of the degrees of freedom will then be as follows:

	Degrees of freedom
Places	5
Varieties	7
Varieties × places	35
Blocks	24
Experimental error	168
Total	239

It should be noted that in this analysis the sums of squares for varieties and for varieties × places are together equal to the total of the sums of squares for varieties in the analyses of the separate experiments. Similarly the sums of squares for blocks and for experimental error are equal to the totals of these items in the separate analyses. If, as is usual, the comprehensive analysis is given in units of a single plot yield, the sums of squares derived from the two-way table of places and varieties must be multiplied or divided by 5 according as means or totals are there tabulated.

The first point to notice about this comprehensive analysis of variance is that the estimates of error from all six places are pooled. If the errors of all experiments are substantially the same, such pooling gives a more accurate estimate than the estimates derived from each separate experiment, since a larger number of degrees of freedom is available. If the errors are different, the pooled estimate of the error variance is in fact the estimate of the mean of the error variances of the separate experiments. It will therefore still be the correct estimate of the error affecting the mean difference (over all places) of two varieties, but it will no longer be applicable to comparisons involving some of the places only. Moreover, as will be explained in more detail below, the ordinary tests of significance, even of means over all places, will be incorrect.

If the errors of all the experiments are the same, the other mean squares in the analysis of variance table may be compared with the mean square for experimental error by means of the z test. The two comparisons of chief interest are those for varieties and for varieties × places. The meaning of these will be clear if we remember that there is a separate set of varietal means at each place, and that the differences between these means are not necessarily the same at all places. If the mean square for varieties is significant, this indicates the significance of the average differences of the varieties *over the particular set of places chosen.* If varieties × places is also significant, a significant variation from place to place in the varietal differences is indicated. In this latter case it is clear that the choice of places must affect the magnitude of the average differences between varieties: with a different set of places we might obtain a substantially different set of average differences. Even if varieties × places is not significant, this fact cannot be taken as indicating *no* variation in the varietal differences, but only that such variation is likely to be smaller than an amount which can be determined by the arguments of fiducial probability.

We may, therefore, desire to determine the variation that is likely to occur in the average differences between the varieties when different sets of places are chosen, and in particular whether the average differences actually obtained differ significantly from zero when variation from place to place is allowed for. Endless complications affect this question, and with the material ordinarily available a definite answer is usually impossible. The various points that arise will be made clear by an actual example, but first we may consider the situation in the ideal case where the chosen places are a strictly random selection from all possible places.

At first sight it would appear to be legitimate in this case to compare the mean square for varieties with that for varieties × places by means of the z test. There is, however, no reason to suppose that the variation of the varietal differences from place to place is the same for each pair of varieties. Thus the eight varieties of our example might consist of two sets of four, the varieties of each set being closely similar among themselves but differing widely from those of the other set, not only in their average yield, but also in their variations in yield from place to place.

The sums of squares for varieties and for varieties × places would then have large components derived from one degree and five degrees of freedom respectively, while the remaining components might be of the order of experimental error. In the limiting case, therefore, when the experimental error is negligible in comparison with the differences of the two sets and the average difference over all possible places is zero, the z derived from the two mean squares will be distributed as z for 1 and 5 degrees of freedom instead of as z for 7 and 35 degrees of freedom. Verdicts of significance and of subnormal variation will therefore be reached far more often than they should be.

The correct procedure in this case is to divide the sums of squares for varieties and for varieties × places into separate components, and compare each component separately. Thus we shall have:

	Degrees of freedom
Varieties: Sets	1
Within sets	6
Varieties × places: Sets	5
Within sets	30

The 1 degree of freedom between sets can now legitimately be compared with the 5 degrees of freedom for sets × places, but the degrees of freedom within sets may require further subdivision before comparison.

It is worth noting that the test of a single degree of freedom can be made by the t test, by tabulating the differences between the means of the two sets for each place separately. This test is in practice often more convenient than the z test, of which it is the equivalent.

4. An actual example of the analysis of a set of variety trials

Table I, which has been reproduced by Fisher (1935) as an example for analysis by the reader, gives the results of twelve variety trials on barley conducted in the State of Minnesota and discussed by Immer *et al.* (1934). The trials were carried out at six experiment stations in each of two years, and actually included ten varieties of which only five (those selected by Fisher) are considered here.

Table I. *Yields of barley varieties in twelve independent trials. Totals of three plots, in bushels per acre*

Place and year	Manchuria	Svansota	Velvet	Trebi	Peatland	Total
University Farm 1931	81·0	105·4	119·7	109·7	98·3	514·1
1932	80·7	82·3	80·4	87·2	84·2	414·8
Waseca 1931	146·6	142·0	150·7	191·5	145·7	776·5
1932	100·4	115·5	112·2	147·7	108·1	583·9
Morris 1931	82·3	77·3	78·4	131·3	89·6	458·9
1932	103·1	105·1	116·5	139·9	129·6	594·2
Crookston 1931	119·8	121·4	124·0	140·8	124·8	630·8
1932	98·9	61·9	96·2	125·5	75·7	458·2
Grand Rapids 1931	98·9	89·0	69·1	89·3	104·1	450·4
1932	66·4	49·9	96·7	61·9	80·3	355·2
Duluth 1931	86·9	77·1	78·9	101·8	96·0	440·7
1932	67·7	66·7	67·4	91·8	94·1	387·7
Total	1132·7	1093·6	1190·2	1418·4	1230·5	6065·4

The experiments were all arranged in randomized blocks with three replicates of each variety. When all ten varieties are included there are therefore 18 degrees of freedom for experimental error at each station.

The error mean squares for the twelve experiments were computed from the yields of the separate plots, which have been given in full by Immer. They are shown in Table II.

Table II. *Error mean squares of barley experiments*

	Mean square		Approximate χ^2	
	1931	1932	1931	1932
University Farm	21·25	15·98	16·43	12·36
Waseca	26·11	25·21	20·19	19·49
Morris	18·62	20·03	14·40	15·49
Crookston	30·27	21·95	23·40	16·97
Grand Rapids	26·28	26·40	20·32	20·41
Duluth	27·00	20·28	20·88	15·68

If the errors at all stations are in fact the same, these error mean squares, when divided by the true error variance and multiplied by the number of degrees of freedom, 18, will be distributed as χ^2 with 18 degrees of freedom. If we take the mean of the error mean squares, 23·28, as an estimate of the true error variance, the distribution obtained will not be far removed from the χ^2 distribution. The actual values so obtained are shown in Table II, and their distribution is compared with the χ^2 distribution in Table III. Variation in the experimental error from station to station would be indicated by this distribution having a wider dispersion than the χ^2 distribution.

Table III. *Comparison with the theoretical χ^2 distribution*

P	1·0	0·99	0·98	0·95	0·90	0·80	0·70	0·50	0·30	0·20	0·10	0·05	0·02	0·01	0
χ^2	0	7·02	7·91	9·39	10·86	12·86	14·44	17·34	20·60	22·76	25·99	28·87	32·35	34·80	∞
No. observed	0	0	0	0	1	1	4	4	1	1	0	0	0	0	
No. expected	0·12	0·12	0·36	0·6	1·2	1·2	2·4	2·4	1·2	1·2	0·36	0·36	0·12	0·12	

In this case it is clear that the agreement is good, and consequently we shall be doing no violence to the data if we assume that the experimental errors are the same for all the experiments. This gives 23·28 as the general estimate (216 degrees of freedom) for the error variance of a single plot, and the standard error of the values of Table I is therefore $\sqrt{(3 \times 23{\cdot}28)}$ or $\pm 8{\cdot}36$.

The analysis of variance of the values of Table I is given in Table IV in units of a single plot. The components due to places and years have been separated in the analysis in the ordinary manner.

Every mean square, except that for varieties × years, will be found, on testing by the z test, to be significantly above the error mean square. Examination of Table I indicates that variety Trebi is accounting for a good deal of the additional variation due to varieties and varieties × places, for the mean yield of this variety over all the experiments is much above that of the other four varieties, but at University Farm and Grand Rapids, two of the lowest yielding stations, it has done no better than the other varieties.

Table IV. *General analysis of variance (units of a single plot)*

	Degrees of freedom	Sum of squares	Mean square	
Places	5	7073·64	1414·73	
Years	1	1266·17	1266·17	
Places × years	5	2297·96	459·59	
Varieties	4	1769·99	442·50	
Varieties × places	20	1477·67	73·88	
Varieties × years	4	97·27	24·32	} 42·72
Varieties × places × years	20	928·09	46·40	
Total	59	14910·79		
Experimental error	216		23·28	

In order to separate the effect of Trebi it is necessary to calculate the difference between the yield of Trebi and the mean of the yields of the other four varieties for each of the twelve experiments, and to analyse the variance of these quantities. For purposes of computation the quantities:

$$5 \times \text{yield of Trebi} - \text{total yield of station} = 4\,(\text{yield of Trebi} - \text{mean of other varieties})$$

are more convenient.

The analysis of variance is similar to that which gives places, years and places × years in the main analysis. The divisor of the square of a single quantity is $3 \times (4^2+1+1+1+1) = 60$ and the square of the total of all twelve quantities, divided by 720, gives the sum of squares representing the average difference between Trebi and the other varieties over all stations.

The four items involving varieties in the analysis of variance are thus each split up into two parts, representing the difference of Trebi and the other varieties, and the variation of these other varieties among themselves. This partition is given in Table V. The second part of each sum of

Table V. *Analysis of variance, Trebi v. Remainder*

	Degrees of freedom	Sum of squares	Mean square
Varieties: Trebi	1	1463·76	1463·76
Remainder	3	306·23	102·08
Varieties × places: Trebi	5	938·09	187·62
Remainder	15	539·58	35·97
Varieties × years: Trebi	1	7·73	7·73
Remainder	3	89·54	29·85
Varieties × places × years: Trebi	5	162·10	32·42
Remainder	15	765·97	51·06

squares can be derived by subtraction, or by calculating the sum of squares of the deviations of the four remaining varieties from their own mean.

Study of this table immediately shows that the majority of the variation in varietal differences between places is accounted for by the difference of Trebi from the other varieties. The mean square for varieties × places has been reduced from 73·88 to 35·97 by the elimination of Trebi, and this latter is in itself not significantly above the experimental error. The mean square for varieties × places × years has not been similarly reduced, however, in fact it is actually increased (though not significantly), and the last three remainder items taken together are still significantly above the experimental error. There is thus still some slight additional variation in response from year to year and place to place. The place to place variation appears to arise about equally from all the remaining three varieties, but the differences between the different years are almost wholly attributable to the anomalous behaviour at Grand Rapids of Velvet, which yielded low in 1931 and high in 1932. If the one degree of freedom from varieties × years and varieties × places × years arising from this difference is eliminated by the "missing plot" technique (Yates, 1933) we have

	Degrees of freedom	Sum of squares	Mean squares
Velvet at Grand Rapids	1	453·19	453·19
Remainder	23	572·16	24·88

Thus the remainder is all accounted for by experimental error.

It has already been noted that Trebi yielded relatively highly at the high-yielding centres. The degree of association between varietal differences and general fertility (as indicated by the mean of all five varieties) can be further investigated by calculating the regressions of the yields of the separate varieties on the mean yields of all varieties. The deviations of the mean yields of the six stations from the general mean in the order given are nearly proportional to

$$-2, +10, +1, +2, -6, -5.$$

The sum of the squares of these numbers is 170. Multiplying the varietal totals at each place by these numbers we obtain the sums

Manchuria	1004·6
Svansota	1196·2
Velvet	1137·8
Trebi	1926·8
Peatland	736·3
Total	6001·7

The sum of the squares of the deviations of these sums, divided by 170×6, gives the part of the sum of squares accounted for by the differences of the regressions. This can be further subdivided as before, giving:

	Degrees of freedom	Sum of squares	Mean square
Varieties × Places:			
Differences of regressions: Trebi	1	646·75	646·75
Remainder	3	123·17	41·06
Total	4	769·92	
Deviations from regressions: Trebi	4	291·34	72·84
Remainder	12	416·39	34·70
Total	16	707·73	

Thus the greater part of the differences between Trebi and the remaining varieties is accounted for by a linear regression on mean yield. There is still a somewhat higher residual variation (M.S. 72·84) of Trebi from its own regression than of the other varieties from their regressions, though the difference is not significant. Of the four remaining varieties Peatland appears to have a lower regression than the others, giving significantly higher yields at the lower yielding stations only, the difference in regressions being significant when Peatland is tested against the other three varieties.

The whole situation is set out in graphical form in Fig. 1, where the stations are arranged according to their mean yields and the calculated regressions are shown. It may be mentioned that of the remaining five varieties not included in the discussion, three show similar regressions on mean yield.

This artifice of taking the regression on the mean yield of the difference of one variety from the mean of the others is frequently of use in revealing relations between general fertility and varietal differences. A similar procedure can be followed with response to fertilizers or other treatments. The object of taking the regression on the mean yield rather than on the yield of the remaining varieties is to eliminate a spurious component of regression which will otherwise be introduced by experimental errors. If the variability of each variety at each centre is the same, apart from the components of variability accounted for by the regressions, the regression so obtained will give a correct impression of the results. This is always the situation as far as experimental error is concerned (except in those rare cases in which one variety is more variable than the others). It may or may not be the case for the other components of variability. In our example, as we have seen, the deviations of Trebi from its regression are somewhat greater than those of the other varieties. In such cases we should theoretically take regressions on a weighted mean yield, but there will be little change in the results unless the additional components of variance are very large.

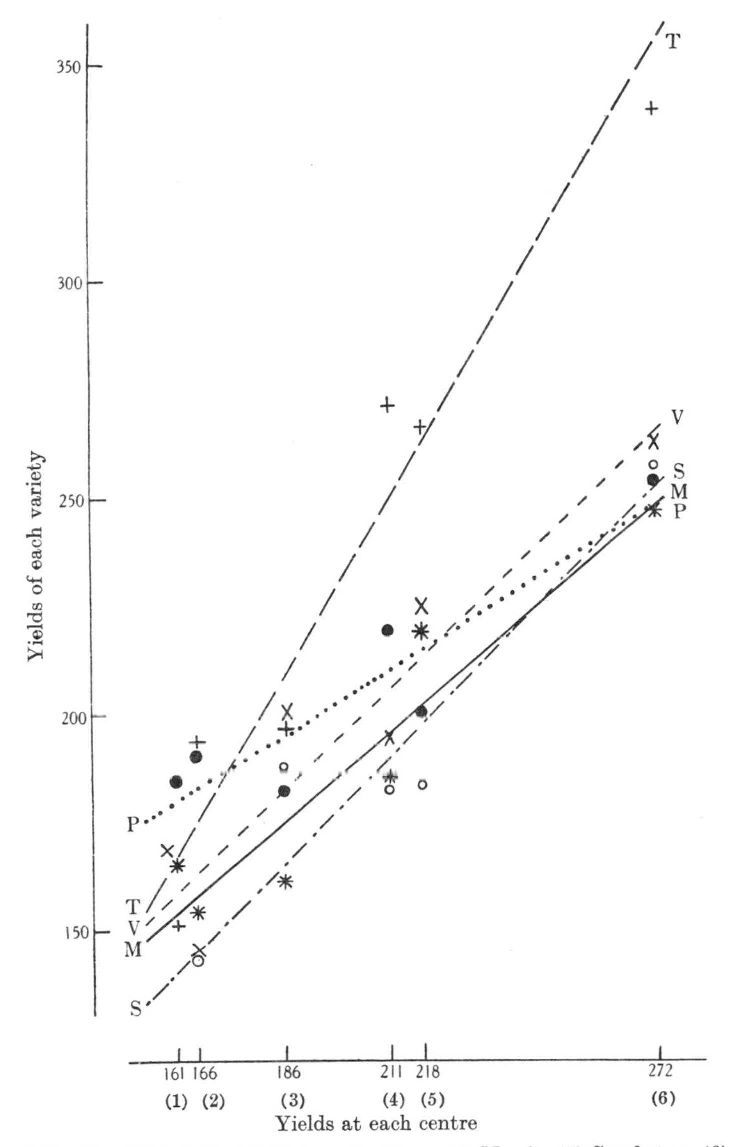

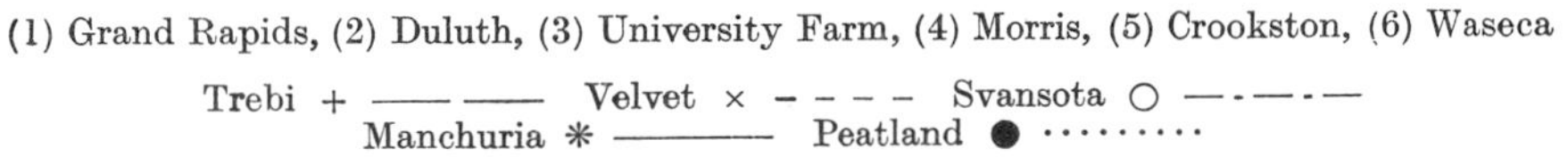
(1) Grand Rapids, (2) Duluth, (3) University Farm, (4) Morris, (5) Crookston, (6) Waseca

Trebi + ——— Velvet × - - - - Svansota ○ —·—·—
Manchuria ✳ ——— Peatland ● ·········

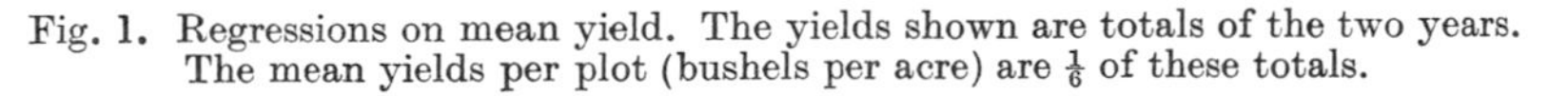
Fig. 1. Regressions on mean yield. The yields shown are totals of the two years. The mean yields per plot (bushels per acre) are $\frac{1}{6}$ of these totals.

We can also examine how far it is possible to make practical recommendations from the results of these experiments.[1] The following points are of importance:

(1) How far is the superiority of Trebi over the remaining four varieties likely to recur in future years at the same set of stations?

(2) How far is Trebi likely to be superior to the other varieties at other stations in years similar to 1931 and 1932, and in particular with what degree of confidence may Trebi be recommended for general adoption in preference to the other four varieties in some or all of the districts of the state of Minnesota?

The answer to question (1) is clearly limited by the fact that only 2 years' results are available. Therefore, we cannot make any general statement as regards years which are radically different in weather conditions to 1931 and 1932. If, however, 1931 and 1932 themselves differed considerably as regards weather conditions, and if, moreover, the weather conditions varied considerably from station to station in the same year, the weather conditions over the twelve experiments might be regarded as an approximation to a random sample from all possible conditions, in which case the pooled estimate of varieties × years and varieties × places × years (these not being significantly different) might be regarded as an appropriate estimate of the variance due to weather conditions, to differences between fields at the same station, and to experimental error. In this respect Trebi is no more variable than the remaining varieties, and if the anomalous variation of Velvet at Grand Rapids be excluded the pooled estimate (23 degrees of freedom) is only slightly above experimental error. In default of any special explanation of this anomalous variation, however, it will be more reasonable not to exclude this degree of freedom, in which case we should assess the variance due to the above causes as 42·72 (24 degrees of freedom) and this agrees closely with the similar estimate (which cannot be attributed to any one outstanding difference) from the varieties which have been omitted from the present analysis.

Since three plots go to make up the total at any one place in one year, the additional variance of a single varietal mean (on a single plot basis) due to weather conditions and differences between fields at the same station, over that arising from experimental error, is

$$\tfrac{1}{3}\,(42{\cdot}72 - 23{\cdot}28) = 6{\cdot}48,$$

[1] It should be noted that in practice there are many other factors besides yield which must be taken into consideration. In this instance Dr Immer informs me that the malting quality of Trebi is poor.

and the variance of the difference of two varieties due to these causes will be double this, i.e. 12·96.

In addition to this real variation in subsequent years of the varietal differences at a place, the errors of the estimates obtained from the 2 years' experimental results must also be taken into account. These are calculated in the ordinary manner from the mean square, 42·72, for varieties × years, and varieties × places × years. Thus the error variance of the estimate of the difference of two varieties at any one place is twice $\frac{1}{6}$ (42·72) = 14·24. (If varieties × years and varieties × places × years were different further subdivision of the components, similar to that illustrated below when considering place to place variation, would be necessary.)

In one sense, therefore, the variance of the expected difference of any two varieties, say of Trebi and Peatland at Waseca, in any subsequent year (with similar weather conditions) is 12·96 + 14·24 = 27·20, but it must be remembered that if a whole series of subsequent years is taken the actual differences will not be distributed about the estimated difference 14·2 with variance 27·20, but about some unknown "true" difference with variance 12·96, the unknown true difference having itself a fiducial distribution about the estimated difference given by the variance 14·24.

The answer to the second question depends on how far the actual stations may be regarded as a random sample of all stations. If this is the case, the estimate of varieties × places for Trebi will be the appropriate estimate of the variance from place to place, including one-half the variance due to differences between fields at the same station, to experimental error and to weather conditions except in so far as they are constant over all stations in each year. This is based on only five degrees of freedom and is therefore ill determined, but accepting the value 187·62, the variance of the difference of Trebi and the remaining varieties due to places only is:

$$\tfrac{1}{6}\,(1+\tfrac{1}{4})\,(187{\cdot}62-42{\cdot}72)=30{\cdot}19,$$

and, therefore, that due to place, field and weather conditions (but excluding experimental error) is

$$30{\cdot}19+(1+\tfrac{1}{4})\,6{\cdot}48=38{\cdot}29.$$

In addition to this variation the error of the estimated mean difference must be taken into account. The mean difference of Trebi from the mean of the other varieties is 7·1 per plot, and this has an estimated variance due to places, fields, differences in weather conditions from place to place, and experimental error, of

$$\tfrac{1}{36}\,(1+\tfrac{1}{4})\,187{\cdot}62=6{\cdot}51,$$

and also an additional undetermined component of variance due to differences between years.

Hence, the difference to be expected in a single field is subject to a variation about the true mean having an estimated variance of 38·29, and the estimate of the true mean 7·1 has an error variance of 6·51. Consequently it will frequently happen in default of other information that Trebi will yield less than some other variety that might have been grown. At Grand Rapids, Trebi yielded 15–20 % less than Peatland in the 2 years of the experiment. It is poor consolation to the farmer of a farm similar to this to be told that Trebi is giving substantially higher yields on other farms.

It would be rash, however, to recommend Peatland for the whole of the Grand Rapids and Duluth districts and Trebi for the whole of the other districts, in particular the Waseca district, until we know how far variation in the varietal differences depends on factors common to the whole of a district or soil type and how far on factors exclusive to individual farms, such as variations in manuring and cultivation practices and differing crop rotations. Only parallel experiments in the same district on farms which may themselves be reasonably regarded as a random selection from all farms in the district will separate these two sources of variation, and it is therefore impossible from the general analysis of variance to say with any confidence whether Trebi is particularly suited to the district of Waseca or whether its high yield here is due to special conditions at the experimental station. As we have seen, however, the superiority of Trebi is associated with high general level of fertility. If, therefore, we know that the Waseca district is as a whole high yielding we may confidently recommend Trebi for general adoption in the district (with a reservation as to weather conditions). On the other hand, if the general yield of the district is only average, the experimental station being outstanding, then we should only be justified in recommending Trebi for farms of high fertility in this district but might also include farms of high fertility in other districts.

Immer does not report the soil types of the various stations, but it is noteworthy that Peatland, which proved the best variety at the low-yielding stations, has (as its name implies) been specially selected for peat soils, which are likely to be low yielding.

5. Example of variation in experimental error

If the experimental errors of the different experiments are substantially different the use of the z test in conjunction with the pooled estimate of error may be misleading, in just the same way as the pooling of all the degrees of freedom from varieties × places was misleading in the set of varietal trials already considered.

The following is an example in which the z test indicates an almost significant interaction between a treatment effect and places, whereas proper tests show that there is no indication of any such variation in the treatment effect. The example is particularly interesting in that on a first examination of the data the results of the z test led the experimenter to draw false conclusions.

The experiments consisted of a series of thirteen 3 × 3 Latin squares, described by Lewis & Trevains (1934) and carried out in order to test the effectiveness of, and difference between, an ammonium phosphate mixture and an ordinary fertilizer mixture on sugar beet. Large responses to the fertilizers were shown in all the experiments. The question arose as to whether there was any significant difference between the two forms of fertilizer.

Table VI. *Experiments on sugar beet*

Centre	Yields of roots, tons per acre: No fertilizer	Amm. phos. mixture	Ordinary mixture	Mean response	Amm. phos. − ordinary (x)	Error mean square per plot (s^2)	χ^2	t
1	7·44	15·69	13·75	+7·28	+1·94	0·8599	3·74	+2·56
2	7·19	12·28	11·32	+4·61	+0·96	0·3543	1·54	+1·98
3	10·07	13·93	13·10	+3·44	+0·83	0·5329	2·32	+1·39
4	7·74	10·97	11·89	+3·69	−0·92	1·1528	5·02	−1·05
5	11·88	13·96	15·06	+2·63	−1·10	0·2638	1·15	−2·68
6	11·94	14·35	14·36	+2·42	−0·01	1·7249	7·51	−0·01
7	6·20	10·27	10·02	+3·94	+0·25	0·4803	2·09	+0·44
8	8·99	11·17	11·47	+2·33	−0·30	0·1107	0·482	−1·10
9	9·46	12·54	12·46	+3·04	+0·08	0·0184	0·0801	+0·72
10	7·42	10·93	10·79	+3·44	+0·14	0·0046	0·0200	+2·53
11	3·70	5·46	5·38	+1·72	+0·08	0·0073	0·0318	+1·15
12	9·62	12·72	13·01	+3·24	−0·29	0·1920	0·836	−0·81
13	9·47	13·53	13·72	+4·16	−0·19	0·2706	1·18	−0·45
Mean	8·55	12·14	12·02	+3·53	+0·1115	0·4594	2·00	

The yields are shown in Table VI. The mean yields of the two forms of fertilizer over all experiments are practically identical, indicating an absence of any consistent difference between the two forms. The analysis of variance, using a pooled estimate of error from all squares, is given in Table VII.

The z between treatment × centres and error is 0·361, which is almost significant, the 5% point being 0·382. Inspection of the differences between the two forms shows that eight of these are small (⩽ 0·30), while the remaining five range numerically from 0·83 to 1·94. These five are all associated with large error mean squares. The values of t for the separate

Table VII. *Analysis of variance of responses to fertilizer and of difference between mixtures (single-plot basis)*

	Degrees of freedom	Sum of squares	Mean square	z
Response to fertilizer:				
Mean response	1	324·7605	324·7605	
Response × centres	12	45·8462	3·8205	
Differences between mixtures:				
Mean difference	1	0·2493	0·2493	
Difference × centres	12	11·3542	0·9462	0·361
Error	26	11·9450	0·4594	

experiments have therefore been calculated and are given in Table VI. These thirteen observed values are compared with the theoretical t distribution for 2 degrees of freedom in Table VIII. The two distributions agree excellently, not one of the values of t being below the 0·1 level of probability. We must conclude, therefore, that there is no evidence from the experiments that the two mixtures behaved differently at any centre.

Table VIII. *Comparison with theoretical t and χ^2 distributions*

P	1·0		0·9		0·8		0·7		0·6		0·5		0·4		0·3		0·2		0·1		0·05		0·02		0·01		0
t	0		0·14		0·29		0·44		0·62		0·82		1·06		1·39		1·89		2·92		4·30		6·96		9·92		∞
Expected		1·3		1·3		1·3		1·3		1·3		1·3		1·3		1·3		1·3		0·65		0·39		0·13		0·13	
Observed		1		0		½		1½		2		1		2½		½		4		0		0		0		0	

P	1·0		0·99		0·98		0·95		0·90		0·80		0·70		0·50		0·30		0·20		0·10		0·05		0·02		0·01		0
χ^2	0		0·0201		0·0404		0·103		0·211		0·446		0·713		1·39		2·41		3·22		4·60		5·99		7·82		9·21		∞
Expected		0·13		0·13		0·39		0·65		1·3		1·3		2·6		2·6		1·3		1·3		0·65		0·39		0·13		0·13	
Observed		1		1		1		0		0		1		3		3		0		1		1		1		0		0	

The above method requires modification if the true difference μ between the two forms of fertilizer is appreciably different from zero, for the quantities $t' = (x-\mu)/S$ will then conform to the t distribution, instead of the t's calculated as above. The quantity μ is not exactly known, but if the centres are at all numerous the use of the mean difference $\bar{x}$, or some form of weighted mean difference, such as one of those discussed in the next section, will give quantities which closely approximate to the t distribution.

Although inspection of Table VIII shows quite conclusively in the present example that the observed t's are in no way abnormal, border-line cases will arise in which a proper test of significance is desirable. An obvious form of test would be that based on the variance of the observed t's, or of some analogous function. One such function is the "weighted sum of squares of deviations",

$$Q = Sw\,(x-\bar{x}_w)^2 = Swx^2 - \bar{x}_w \,.\, Swx,$$

where the weights w are the reciprocals of the estimates of the error variances of the differences x, and $\bar{x}_w$ is the weighted mean of the x's, i.e.

$$\bar{x}_w = \frac{Swx}{Sw}.$$

If we then calculate

$$\chi'^2 = (k-1) + \sqrt{\frac{n-4}{n-1}} \left\{ \frac{n-2}{n} Q - (k-1) \right\},$$

χ'^2 will be distributed approximately as χ^2 with $k-1$ degrees of freedom. The relation of Q to the ordinary expression for the variance of t' is shown by the alternative form

$$Q = S(t'^2) - \{S(t'\sqrt{w})\}^2 / S(w).$$

This test should not be used if n is less than 6. Actual comparison of the distribution of the t''s with the t distribution should then be resorted to, the unweighted mean $\bar{x}$ being used as the estimate of μ.

An example of the calculation of Q and χ' is given in § 7.

6. Methods of estimating the average response

As has been pointed out in the second section, the average response to a treatment over a set of experiments is frequently of considerable importance, even when the response varies from experiment to experiment. The problem of how it may best be estimated from the results of the separate experiments must therefore be considered.

If the experiments are all of equal precision the efficient estimate is clearly the ordinary mean of the apparent responses in each experiment, whether the true responses are the same or vary from experiment to experiment. If, on the other hand, some of the experiments are more precise than others, the ordinary mean, by giving equal weight to both the less and the more accurate results, may appear at first sight to furnish a considerably less precise estimate than might be obtained by more refined statistical processes. As will appear from what follows, however, there are several factors which increase the advantages of the ordinary mean in relation to other possible estimates, so that unless the experiments differ widely in accuracy, or the conditions are somewhat different from those ordinarily met with in agriculture, the ordinary mean is in practice the most satisfactory as well as the most straightforward estimate to adopt.

The simplest alternative to the ordinary mean is the weighted mean mentioned at the end of the last section, in which the weights are inversely proportional to the error variances of the estimates derived from the various experiments. This weighted mean would be the efficient estimate if there were no variation in the true response from experiment to experiment, and if, moreover, the error variances of the experiments were accurately known. If the error variances are only estimated from a small number of degrees of freedom, however, the weighted mean loses greatly in efficiency and is frequently less efficient than the unweighted mean.

If the true response varies from experiment to experiment, having a variance of σ_0^2, and the error variances are accurately known, the efficient estimate of the mean response is provided by a weighted mean with weights inversely proportional to $\sigma_0^2+\sigma_1^2$, $\sigma_0^2+\sigma_2^2\ldots$, where σ_1^2, $\sigma_2^2,\ldots$, are the error variances of the estimates from the various experiments. This has been called the *semi-weighted mean*, since the weights are intermediate between those of the weighted mean and the equal weights of the ordinary mean.

If the response does not vary from experiment to experiment, but the error variances are not accurately known, being estimated from n_1, n_2, ..., degrees of freedom, the efficient estimate is obtained by the solution of the maximum likelihood equation:

$$\mu = S\,\frac{(n_i+1)\,x_i}{n_i s_i^2+(x_i-\mu)^2}\Big/ S\,\frac{(n_i+1)}{n_i s_i^2+(x_i-\mu)^2}.$$

This solution has the effect of giving lower weights to the more discrepant values than would be given by the ordinary weighted mean. It is not difficult to solve the equation by successive approximation, starting with a value of μ equal to the unweighted mean, but since in agricultural experiments cases in which the response can confidently be asserted not to vary are rare, the additional numerical work is not ordinarily justifiable, except when exact tests of a significance are required and when the n_i are small.

Thus the available rigorous methods of weighting are not of much use in the reduction of the results of the type ordinarily met with. On the other hand, when a set of experiments of widely varying accuracy is encountered, some method of discounting the results of the less accurate experiments is required. The simplest method would be to reject the results of the less accurate experiments entirely, but this involves the drawing of an arbitrary line of division. Anyone who has attempted this will know how easy it is in certain cases to produce substantial changes in the mean response by the inclusion or exclusion of certain border-line experiments.

An alternative procedure is that of fixing an upper limit to the weight assignable to any one experiment. All experiments having error variances which give apparent weights greater than this upper limit are treated as of equal weight. Experiments having a lesser accuracy are weighted inversely as their error variances. The efficiency of this procedure is discussed in Cochran (1937), where it is shown to be substantially more efficient than the use of the ordinary weighted mean if the numbers of degrees of freedom for error are small. Quite large changes in the choice of the upper limit do not seriously affect the efficiency, and equally will not produce any great changes in the resultant estimate. In most agricultural field experiments the upper limit given by an error variance corresponding to a standard error of 5–7 % per plot would seem appropriate in cases in which there is no evidence of variation in the response from experiment to experiment.

A further alternative procedure which produces much the same effect is provided by the use of the semi-weighted mean, assigning some arbitrary value to σ_0^2. This procedure has the advantage of being easily adaptable to cases in which there is evidence of variation in response from experiment to experiment. If, for instance, there are eight replicates of each treatment, the value of σ_0^2 corresponding to 4 % per plot will be $(0{\cdot}04)^2\,(\frac{1}{8}+\frac{1}{8})$, i.e. 0·0004 times the square of the mean yield. This will produce about the same effect as taking a lower limit corresponding to a standard error of 5 % per plot. If in addition the estimated variance of the response from centre to centre is 0·0006 times the square of the mean yield, then we might reasonably take a value of σ_0^2 corresponding to 0·0010 times the square of the mean yield.

If the error variances of the various experiments are accurately known the error variance of any form of weighted mean is given by

$$\frac{w_1'^2\,\sigma_1^2+w_2'^2\,\sigma_2^2+\ldots}{(w_1'+w_2'+\ldots)^2},$$

where w_1', w_2', ... represent the weights actually adopted. If w_1', w_2', ... are equal to $1/\sigma_1^2$, $1/\sigma_2^2$, ..., this expression reduces to the expression for the error variance of the fully weighted mean, namely $1/(w_1'+w_2'+\ldots)$, and if all the weights are equal the error variance of the unweighted mean of k estimates

$$\frac{1}{k^2}\,(\sigma_1^2+\sigma_2^2+\ldots)$$

is obtained.

If, however, the error variances are estimated, and the weights depend on these estimates, the above expression will not be correct. In particular the estimated error variance of the fully weighted mean in a group of experiments each with n degrees of freedom for error will be $n/(n-4)$ times the expression given above (the variances being replaced by their estimates throughout). The error variance of any semi-weighted mean, or weighted mean with upper limit, will have to be similarly increased. No exact expressions are available, but in general the additional factor must lie between $n/(n-4)$ and unity. In the case of the weighted mean with upper limit to the weights the inclusion of the factor $n/(n-4)$ in the terms which have weights below the upper limit is likely to give a reasonable approximation.

The mean (weighted or unweighted) may be tested for significance by means of the t test, using the estimated standard error. The test is not exact, since the number of degrees of freedom is not properly defined, but if a number somewhat less than the total number of degrees of freedom for error in the whole set of experiments is chosen the test will be quite satisfactory.

There is one further point that must be examined before using any form of mean in which the weights depend on the relative precision of the various experiments. If the precision is associated in any way with the magnitude of the response, such a weighted mean will produce biased estimates and must not be used. Thus, for example, the response to a fertilizer might be greater on poor land, and this land might be more irregular than good land, so that experiments on poor land would give results of lower precision. In such a case any of the above weighted means would lead to an estimate of the average response which would be smaller than it should be.

To see whether association of this type exists the experiments may be divided into two or more classes according to accuracy, and the differences between the mean response in each class examined. Alternatively the regression of the responses on the standard errors of the experiments may be calculated.

7. EXAMPLE OF THE ANALYSIS OF A SET OF EXPERIMENTS OF UNEQUAL PRECISION

Table IX gives the responses (in yield of roots) to the three standard fertilizers in a set of $3 \times 3 \times 3$ experiments on sugar beet. These experiments were conducted in various beet growing districts in England. The results shown are those of the year 1934 and are reported in full in the Rothamsted Report (1934). It cannot be claimed that the sites were selected at random (practical considerations precluded this course) and consequently any values obtained for the average responses must be accepted with caution, but the results will serve to illustrate the statistical points involved.

Table IX. *Responses to fertilizers in a series of experiments on sugar beet*

Washed roots (tons per acre)

Station	Mean yield	Linear response to N	P	K	Standard error	Wt.	Degrees of freedom
Allscott	10·97	−0·24	+0·63	+0·57	±0·519	3·7	15
Bardney	11·44	+1·23†	+0·35	+0·01	±0·285	12·3	22
Brigg	13·42	+0·11	−0·38	−0·21	±0·603	2·8	22
Bury	13·83	+2·08†	−0·05	−0·22	±0·351		15
Cantley	12·90	+0·20	+0·32	+0·14	±0·453	4·9	15
Colwick	10·12	+1·05†	+0·87†	−0·07	±0·287	12·2	15
Ely	12·46	−1·14	+0·80	−0·08	±0·886	1·3	15
Felstead	11·28	+3·34†	+0·11	+0·23	±0·356	7·9	15
Ipswich	12·45	+1·64†	+0·57	+0·34	±0·344	8·5	15
King's Lynn	19·54	+0·52	+0·12	−0·57	±0·481	4·3	15
Newark	14·10	+1·37†	+0·54*	−0·33	±0·198	25·5	15
Oaklands	12·84	0·00	−0·14	+0·40	±0·622	2·6	15
Peterborough	17·99	−0·14	+1·02	−1·34*	±0·618	2·6	15
Poppleton	14·21	+2·72†	−0·21	−0·18	±0·357	7·8	22
Wissington	14·55	+3·32†	+0·19	+0·38	±0·443	5·1	15
Mean	13·47	+1·07	+0·32	−0·06	±0·125	109·6	246

* 5% significance. † 1% significance.

At Bardney, Brigg and Poppleton there were two complete replications, i.e. fifty-four plots, while at each of the remaining centres there was a single replication, twenty-seven plots, only, the error being estimated from the interactions of the quadratic components of the responses and from the unconfounded second order interactions. At all centres the experiments were arranged in blocks of nine plots.

The size of the plot varied. It is immediately apparent, from inspection, or by application of the process described in § 4, that the experiments are of very varying precision. In general the larger plots, as might be expected, gave the more accurate results, though the gain in precision was not proportional to the increase in area.

(*a*) *The response to nitrogen.*

The response to nitrogen clearly varies significantly from centre to centre, this variation being large in comparison with experimental error. The ordinary analysis of variance of these responses is given in Table X.

The pooled estimate of error is equal to the mean of the squares of the standard errors given in Table IX. The estimate of the standard error of the average response is therefore

$$\sqrt{(0{\cdot}2345/15)} = \pm 0{\cdot}125.$$

Table X. *Analysis of variance of response to nitrogen*

	Degrees of freedom	Sum of squares	Mean square
Average response	1	17·1949	17·1949
Response × centres	14	25·5891	1·8278
Pooled estimate of error			0·2345

Since the errors are unequal the t distribution will not be exactly followed, the actual 5 % point being subject to slight uncertainty, but in any case intermediate between those given by t for 15 and for 246 degrees of freedom.

The variation of the response from centre to centre has therefore an estimated variance of

$$1{\cdot}8278 - 0{\cdot}2345 = 1{\cdot}5933,$$

excluding variance due to error. This method of estimation is not fully efficient, but may be used in cases such as the present in which the variation is large in comparison with the experimental errors.

It is clear that even were the precision of the experiments known with exactitude, the standard errors being those of Table IX, the semi-weighted mean of § 6, which could then be used, would differ little from the unweighted mean, since the weights would only range from

$$\frac{1}{1{\cdot}5933 + 0{\cdot}0392} \text{ to } \frac{1}{1{\cdot}5933 + 0{\cdot}7850},$$

i.e. from 0·61 to 0·42. The unweighted mean is therefore the only estimate of the average response to nitrogen that need be considered. It may be noted that in this set of experiments there appears to be some association between degree of accuracy and magnitude of response to nitrogen. This is an additional reason for not using any form of weighted mean.

(*b*) *The response to superphosphate.*

The responses to superphosphate are of much smaller magnitude than those to nitrogen. Only two, those of Colwick and Newark, are significant, but eleven out of the fifteen are positive, and consequently there is some evidence for a general response.

The unweighted mean of the responses is $+0{\cdot}32$ and the standard error of the quantity is, as before, $\pm 0{\cdot}125$. The unweighted mean is therefore significant.

The weights corresponding to the estimated standard errors are given in Table IX. The sum of the products of these weights and the responses to phosphate, divided by the sum of the weights, gives the weighted mean, $+0{\cdot}365$. This differs somewhat from the unweighted mean, and inspection shows that the difference is largely due to the fact that the two stations which gave significant results received high weights. The weight assigned to Newark is nearly $\frac{1}{4}$ of the total weight. The estimated error of this experiment is 3 % per plot, which would appear to be lower than is likely to be attained in practice. Fixing a lower limit of error at 5 % per plot, which is equivalent to a weight of 10·0 for the experiments of twenty-seven plots and of 20·0 to experiments with fifty-four plots, we obtain the weighted mean with upper limit, 0·323.

Following the rule given in § 6, the estimated standard error of the weighted mean will be given by the square root of

$$\frac{3\cdot 7 \times \frac{15}{11} + 12\cdot 3 \times \frac{22}{18} + \dots}{(3\cdot 7 + 12\cdot 3 + \dots)^2}.$$

This gives the value $\pm 0\cdot 111$. Similarly the estimated standard error of the weighted mean with upper limit to the weights is given by the square root of

$$\frac{3\cdot 7 \times \frac{15}{11} + 12\cdot 3 \times \frac{22}{18} + \dots + \frac{10^2}{12\cdot 2} + \dots}{(3\cdot 7 + 12\cdot 3 + \dots + 10 + \dots)^2}.$$

This gives a value of $\pm 0\cdot 112$. This is presumably somewhat of an over-estimate, as this mean is likely to be somewhat more accurate than the weighted mean.

In order to test whether there is any evidence of variation in the phosphate response from centre to centre the weighted sum of squares of deviations Q may be calculated by the formula given at the end of § 5. For convenience in this calculation it is best to tabulate the products wx of the weights and the responses separately for each centre. The values obtained are

$$Swx^2 = 27\cdot 6068$$
$$\bar{x}_w \, Swx = 14\cdot 6044$$
$$Q = 13\cdot 0024$$

The value of χ'^2 may now be calculated. In the present case the number of degrees of freedom for error varies from experiment to experiment. We will take n to be equal to the mean 16·4 of these numbers. This procedure will be satisfactory if the numbers do not differ too widely and are reasonably large in all experiments. Using this value we have

$$\chi'^2 = 14 + \sqrt{\frac{12\cdot 4}{15\cdot 4}}\left(\frac{14\cdot 4}{16\cdot 4}\, 13\cdot 0024 - 14\right) = 11\cdot 7.$$

Clearly there is no evidence of any variation in response from centre to centre.

It may, however, be considered that although there is no evidence of variation in response, such variation should not be precluded, and that consequently the upper limit of the weights should be lower than the values of 10 and 20 taken above. Fixing the limit at 7·8, so as to give seven of the fifteen experiments equal weight, we obtain a mean of +0·301, with an estimated standard error of ±0·109. In fertilizer experiments such as the present, where from the nature of the treatments, constancy of response would seem unlikely, this last estimate of the mean response appears to be the most satisfactory, since it gives equal weight to all the more accurate experiments and at the same time prevents the less accurate experiments from unduly influencing the results.

(c) *The response to potash.*

The effect of potash shows no significance, either in mean response or in variation in response from centre to centre. The significant depression at Peterborough can consequently be reasonably attributed to chance.

The analysis follows the lines already given and need not be set out in detail here. The weighted mean with upper limit 7·8 has the value −0·03 ± 0·109. The value of χ'^2 for testing the significance of the variation in response is 11·6.

The results discussed here are, of course, only a part of the full results of the experiments. No consideration has been given to the curvature of the response curves, or to the interactions of the different fertilizers. The whole set of experiments provides an excellent illustration of the power of factorial design to provide accurate and comprehensive information. It will be noted, among other things, that the mean responses to the three fertilizers are determined with a standard error of less than 1% of the mean yield.

SUMMARY

When a set of experiments involving the same or similar treatments is carried out at a number of places, or in a number of years, the results usually require comprehensive examination and summary. In general, each set of results must be considered on its merits, and it is not possible to lay down rules of procedure that will be applicable in all cases, but there are certain preliminary steps in the analysis which can be dealt with in general terms. These are discussed in the present paper and illustrated by actual examples. It is pointed out that the ordinary analysis of variance procedure suitable for dealing with the results of a single experiment may require modification, owing to lack of equality in the errors of the different experiments, and owing to non-homogeneity of the components of the interaction of treatments with places and times.

REFERENCES

COCHRAN, W. G. (1937). *J. R. statist. Soc.*, Suppl. **4**, 102–18.

FISHER, R. A. (1935). *The Design of Experiments.* Edinburgh.

IMMER, F. R., HAYES, H. K. & POWERS, LE ROY (1934). *J. Amer. Soc. Agron.* **26**, 403–19.

LEWIS, A. H. & TREVAINS, D. (1934). *Emp. J. exp. Agric.* **2**, 244.

Rep. Rothamst. exp. Sta. (1934), p. 222.

YATES, F. (1933). *Emp. J. exp. Agric.* **1**, 129–42.

Paper VI

A NEW METHOD OF ARRANGING VARIETY TRIALS INVOLVING A LARGE NUMBER OF VARIETIES

From THE JOURNAL OF AGRICULTURAL SCIENCE
Volume XXVI, Part III, pp. 424–455, 1936

Author's Note

This paper introduces the idea of pseudo-factorial (later renamed quasi-factorial) designs. In these designs a factorial system is imposed on a set of treatments, such as varieties, which do not themselves possess a factorial structure. Confounding of interactions between the pseudo-factors then greatly reduces block size. Alternative methods of eliminating soil heterogeneity, namely the use of systematically arranged controls and randomly assigned controls within small blocks, are also discussed.

Quasi-factorial designs have become popular with plant breeders, though they have certain disadvantages which were not appreciated when the paper was written. The most important is that if certain varieties fail or are discarded before harvest, the design becomes less effective. There is the secondary disadvantage that unless the design is balanced, so that all comparisons are of equal accuracy, comparisons of groups of varieties which happen to coincide with the factorial grouping will be particularly inaccurate. Moreover if there is little reduction in error by the use of small blocks, the efficiency of a quasi-factorial design when analysed by the method given in this paper is less than the corresponding randomized block arrangement. This defect is overcome by the recovery of interblock information (Paper VIII).

Subsequent to Paper VI, M. S. Bartlett investigated a suggestion by J. S. Papadakis for adjusting the yields by covariance on the residuals of neighbouring plots (*Journal of Agricultural Science*, **28**, 418-427, 1938). A further investigation of the unidimensional case, when the errors form a first-order autoregressive series, was recently made by A. C. Atkinson (*Biometrika*, **56**, 33-41, 1969). Atkinson showed that the Papadakis estimate is a close approximation to the maximum likelihood estimate, but the estimates of the variance of the treatment contrasts which he obtained from the residual sums of squares were considerably in excess of the true variance (about 20% too high in his example with an autoregressive correlation $\rho = 0{\cdot}8$). The design used, however, was a rather special one, and the 'blocks' term, which is certainly relevant in randomized block experiments, does not appear to have been separated when estimating the residual sums of squares. The method clearly introduces theoretical complications, as Bartlett stressed, and has never been much used because of the tedious computation required. There is also one practical defect not mentioned by Bartlett; the effect of competition between varieties or other treatments will be exaggerated, by a factor $(1 + \frac{1}{2}b)$ when the plots are long and narrow and two neighbouring plots are taken, and by a factor $(1 + \frac{1}{4}b)$ when four neighbouring plots are taken, b being the regression coefficient. Nevertheless, now that computers are available the method seems worthy of further examination.

1969 FRANK YATES

A NEW METHOD OF ARRANGING VARIETY TRIALS INVOLVING A LARGE NUMBER OF VARIETIES

By F. YATES

(*Rothamsted Experimental Station, Harpenden, Herts*)

(With Two Text-figures)

I. Introduction

When a large number of varieties are to be compared in an agricultural field trial the experimenter is confronted with a difficult problem of arrangement. A Latin square containing the whole of the varieties is impossible, being both too unwieldy and requiring too many replications, and even randomised blocks containing the whole of the varieties are unsatisfactory, for the blocks are likely to contain too many plots to eliminate fertility differences efficiently.

A simple method of keeping the block size small is to select one or more varieties as controls and to divide the rest into sets, each set being arranged with the controls in a number of randomised blocks. Unfortunately this method has the disadvantage that comparisons between varieties in different sets are of lower accuracy than comparisons between varieties in the same set. There is the further disadvantage that a disproportionate number of plots is devoted to the control varieties, thus further lowering the efficiency.

The control method can, of course, also be used in conjunction with the Latin square method of design, each set of varieties together with the control or controls being arranged in the form of a Latin square.

Various systematic arrangements of control plots are also possible, the most widely used being that in which the control plot is repeated at fixed intervals along a line of plots. Most frequently in such arrangements a systematic arrangement of the new varieties is also adopted. This is wholly to be condemned, for not only is no valid estimate of error possible, but there is also the disadvantage that differences between neighbouring varieties are likely to be more accurately determined than those between widely separated varieties; moreover, there is a danger of serious disturbance from anything of the nature of a fertility wave. If, however, the new varieties are arranged at random all these objections are overcome, so far as the new varieties are concerned; only the comparisons with the controls are now liable to bias and lacking in an estimate of error.

Whatever the exact lay-out of the controls, any method using them suffers from the disadvantage that part of the land has to be devoted to them which would otherwise be devoted to experimental varieties, thus tending to lower the efficiency of the experiment.

To overcome this defect, and at the same time to avoid the use of excessively large blocks, the *pseudo-factorial* type of arrangement, described in this paper, is suggested. In this type of arrangement the varieties are divided into sets for comparison in more than one way, the sets of each division being so arranged that they cut across those of all the other divisions. Thus 100 varieties, numbered 00–99, may be divided into sets of 10 in two ways, the first group of 10 sets consisting of varieties 00–09, 10–19, 20–29, etc., and the second group of 10 sets consisting of varieties 00, 10, 20, ..., 90; 01, 11, 21, ..., 91; 02, 12, 22, ..., 92; etc. Each set of 10 can then be arranged in the field in the form of one or more randomised blocks of 10 plots each, or in the form of a 10×10 Latin square, according to the number of replications that are feasible.

Information on the difference of two varieties in the same set, such as varieties 00 and 01, will then accrue from two sources. Firstly there will be the direct comparison within the given set 00–09, and secondly 00 may be compared with the mean of 10, 20, ..., 90, and 01 may be compared with the mean of 11, 21, ..., 91 in the sets of the second group, the difference between the two means being determined from comparisons within the first group of sets. The difference between two varieties such as 00 and 11 not occurring in the same set may also be determined in various ways.

More elaborate arrangements of this type are clearly possible. Instead of two groups of sets three may be employed. With 125 varieties, for instance, three groups of 25 sets of 5 may be used, the first group of 25 sets being made up of varieties 1–5, 6–10, 11–15, ..., 121–125, the second group of varieties 1, 6, 11, 16, 21; 2, 7, 12, 17, 22; ..., 26, 31, 36, 41, 46; ..., and the third group of varieties 1, 26, 51, 76, 101; 2, 27, 52, 77, 102; .... With these three groups of sets, using 5×5 Latin squares, 15 replications would be required.

Moreover, the sets of the different groups need not necessarily be of the same size. Thus 90 varieties might be arranged in one group of 9 sets of 10, these being made up of varieties 1–10, 11–20, ..., 81–90, and one group of 10 sets of 9, made up of 1, 11, 21, ..., 81; 2, 12, 22, ..., 82; .... Similar arrangements are possible with three or more groups of sets.

The object of this paper is to discuss in detail the analysis and efficiency of these different types of pseudo-factorial arrangement. As a preliminary the efficiency of arrangements employing controls will be briefly considered.

II. Controls in conjunction with randomised blocks and Latin squares

We will first consider the use of controls in conjunction with ordinary randomised blocks or Latin squares.

Denote the number of plots per block by k, the number of controls per block by c, the number of varieties other than controls by v, the number of replicates of each variety other than controls by r, and the error variance per plot by σ^2.

The number of sets of varieties is then $v/(k-c)$, and the total number of plots N is $rkv/(k-c)$.

The error variances of the various types of comparison are as follows:

Comparison between	Error variance
Two varieties (not controls) in one set	$2\sigma^2/r$
Two varieties not in the same set	$\frac{2\sigma^2}{r}\left(1+\frac{1}{c}\right)$
Mean of controls and one other variety	$\frac{\sigma^2}{r}\left(1+\frac{1}{c}\right)$
One control and one other variety	$\frac{\sigma^2}{r}\left(1+\frac{1}{c}\right)+\frac{\sigma^2}{r}\frac{k-c}{v}\left(1-\frac{1}{c}\right)$

Since all the comparisons have not the same error variance some criterion for assessing the accuracy of the experiment as a whole will be required. The following suggest themselves:

(1) The greatest error variance of any comparison.

(2) The mean error variance of all comparisons.

(3) The mean of the information on all comparisons, *i.e.* the mean of the reciprocals of the error variances of all comparisons.

Criterion (3), which at first sight appears the logical one, does not, however, take into account the advantages that accrue from having all the comparisons of equal accuracy, and for some purposes (1) might be considered the best single criterion. Criterion (2) has the advantage that it has in general a simpler algebraic expression than (3). It will be somewhat greater than (3), though only very slightly so when the variation in accuracy is only small. In what follows, therefore, we shall limit ourselves to the first two criteria, basing our general conclusions on the second.

To determine the optimum number of controls with a given number of plots per block we shall require to determine the minimum of whatever criterion we consider appropriate. We shall first take as a criterion the greatest error variance of any comparison, namely

$$\frac{2\sigma^2}{r}\left(1+\frac{1}{c}\right).$$

This is equal to

$$\frac{2\sigma^2 vk\,(c+1)}{N\,(k-c)\,c}.$$

In the case in which the controls are standard varieties we require a value of c which makes

$$\text{const.}\times\frac{(c+1)}{(k-c)\,c}$$

a minimum. This is given by

$$c\,(k-c)+(c+1)\,(2c-k)=0,$$

$$c=-1+\sqrt{1+k}.$$

The mean error variance of all comparisons, including controls, is

$$V_m=\frac{2\sigma^2 k}{N\,(v+c)\,(v+c-1)}\left[\frac{v^2}{c\,(k-c)}\{(v+c)\,(c+1)-k\}+(v+c)\,(c-1)\right].$$

If the controls are themselves new varieties, and if the mean error variance of all comparisons is chosen as the criterion, the equations for c are more complicated, but will not lead to any widely different values when the number of varieties is large, for comparisons between varieties not in the same set then predominate. In practice only integral values of c are possible, and after calculating c from the above equation the relative mean error variances may be calculated for the two nearest integral values of c in order to decide on which value to adopt.

The efficiency of the arrangement clearly not only depends on the number of controls per block but also on the choice of block size k, hitherto taken as fixed. The error variance per plot will be some function of the block size, and until this function is known it is not possible to determine the most efficient block size. In general, however, it will be found that so far as arrangements in randomised blocks are concerned the method is unlikely to be much more efficient than randomised blocks including all varieties, and on uniform land may be much less efficient. Since it is always likely to be less efficient than the use of systematic controls and pseudo-factorial arrangements there is no need to investigate the matter in more detail here. It need only be noted that the use of randomly placed controls may make possible the employment of Latin squares where these would otherwise be impossible, and for this reason may possess a certain limited usefulness.

In the experiments at Rothamsted and its associated centres, for example, it was found that if no restrictions had been imposed the error variance of the 5×5 Latin squares carried out between 1927 and 1934 would have been increased on the average in the ratio 2·49 : 1. If we were conducting an experiment on 25 varieties we might take one as a control and divide the rest into six sets of four, using six 5×5 Latin squares in all. We should then have $V_m = 0{\cdot}72\sigma^2$. The corresponding variance for randomised blocks of 25 plots would be $V_B = \frac{1}{3}\sigma'^2$. This gives

$$V_B : V_m = 0{\cdot}462\sigma'^2/\sigma^2 = 1{\cdot}15.$$

Thus the method of controls is 15 per cent. more efficient than that of randomised blocks.

There is nothing in the pseudo-factorial type of arrangement which prevents the use of Latin squares instead of randomised blocks, but a greater number of replicates will be required, and this may be impracticable. In the above example a 5×5 pseudo-factorial arrangement would require ten 5×5 Latin squares. Should this number of replicates be possible, however, the pseudo-factorial arrangement (as will be shown later in the paper) would give an efficiency of $\frac{4}{6} \times 2{\cdot}49 = 1{\cdot}66$, and is thus a much better arrangement than the method of controls.

III. Systematic arrangement of controls

One of the disadvantages of the use of controls in conjunction with ordinary randomised blocks is that each control plot is used to furnish information only on the fertility of the block in which it occurs, though it may, for example, fall at the edge of that block and be in fact just as good an indicator of the fertility of plots in its neighbourhood belonging to the next block.

Many methods of utilising the information of systematically arranged control plots have been advocated from time to time, some of them very complicated, and most of them open to serious statistical objections. If, however, the control plots are regarded merely as indicators of the fertility in their neighbourhood and a proper randomisation process is applied to the varieties allotted to all the remaining plots, a statistically valid arrangement for the comparison of the varieties other than controls will result, and appropriate methods of analysis are available.

The simplest type of arrangement of this nature is that in which the plots are arranged in one or more lines, the controls being placed at equal intervals along the lines. Thus, for example, 12 new varieties might be compared in three randomised blocks with controls at every 5th plot. The following arrangement is of this type:

Block I

$$C_1 \quad a \quad l \quad b \quad k \quad C_2 \quad g \quad e \quad f \quad h \quad C_3 \quad c \quad d \quad j \quad i \quad C_4$$

Block II

$$C_5 \quad j \quad g \quad l \quad k \quad C_6 \quad c \quad e \quad d \quad b \quad C_7 \quad a \quad f \quad i \quad h \quad C_8$$

Block III

$$C_9 \quad k \quad c \quad g \quad a \quad C_{10} \quad i \quad l \quad j \quad b \quad C_{11} \quad f \quad e \quad k \quad d \quad C_{12}$$

In analysing such an experiment the experimenter is at liberty to use any function of the controls that he fancies as a measure of the fertility of the neighbouring plots. For practical reasons he will be well advised to choose a simple function, for the advantages accruing from the more complicated functions that have sometimes been suggested are likely to be more theoretical than real.

The simplest function is the mean of the two control plots between which the plot under consideration is situated, but the weighted mean of the two controls, with weights inversely proportional to their distances, would appear to be more suitable. Thus the appropriate values for the first four experimental plots in the above arrangement (varieties a, l, b, k respectively) would be

$$\tfrac{4}{5}C_1+\tfrac{1}{5}C_2,\ \tfrac{3}{5}C_1+\tfrac{2}{5}C_2,\ \tfrac{2}{5}C_1+\tfrac{3}{5}C_2,\ \tfrac{1}{5}C_1+\tfrac{4}{5}C_2.$$

Having decided on the measure of fertility, the experimenter is equally at liberty to use this measure in whatever way appears desirable. He may take the differences between it and the yields of the experimental plots, or he may express the experimental yields as percentages of the corresponding fertility measures, but here again he will be well advised to adopt some simple procedure; in particular percentages are unlikely to possess any advantages over differences. When the corrected values have been obtained it is only necessary to analyse them in the same manner as the results of an ordinary randomised block experiment.

There is, however, a more effective way of using these fertility measures. Such measures are of the nature of concomitant observations, and can be treated in the same manner as other information of this type, such as the yields of a preliminary uniformity trial, namely, by the procedure of the analysis of covariance. The corrected values will then be of the form $y_u - bf_u$, where y_u is the uncorrected yield of the u'th plot, f_u the fertility measure, and b is a constant whose value is determined from the observations themselves, being such that the error of the corrected yields is minimised. By this procedure the danger of overcorrection for the fertility measures, which will almost certainly occur if straight differences are taken, will be avoided, and only that weight will be attached to the fertility measures which is justified by the results of the particular experiment under consideration.

In cases where the majority of the variation in plot yields is due to causes other than fertility differences, or where the control plots provide an inadequate representation of such differences, a value of b much smaller than unity may be anticipated, and the covariance method will then be markedly the more accurate. In the limiting case where there is no association between the yields of neighbouring plots, so that the plot errors may be regarded as independent, with, say, a variance σ^2, the variances of the fertility indices of each set of four plots (with controls at intervals of five plots) are 17/25, 13/25, 13/25 and 17/25 times σ^2 respectively, the mean being $\frac{3}{5}\sigma^2$. The covariance will in this case be zero, and therefore the variance of the differences $y_u - f_u$ will be $\frac{8}{5}\sigma^2$. (With controls at intervals of n plots this variance equals $(5n-1)\,\sigma^2/3n$.) The method of covariance will give b zero, and we shall therefore be using the unaltered yields, which will have a variance of σ^2. The loss of efficiency in this case by the use of the differences $y_u - f_u$ in place of $y_u - bf_u$ is therefore very considerable, ranging from $\frac{1}{3}$ when every alternate plot is a control to $\frac{2}{5}$ when the interval between controls is large.

Apart from the avoidable loss of information due to the use of differences $y_u - f_u$, or other arbitrary functions, in place of $y_u - bf_u$, allowance must also be made, when comparing the efficiency of systematic controls with ordinary randomised blocks, for the reduction in the number of plots available for the experimental varieties due to the existence of the controls. The ratio of the amounts of information tends in the limiting case just considered to the ratio of the number of non-control plots to the total number of plots.

There is, of course, nothing in the method of controls that precludes the use of a fertility index deduced from controls of more than one line of plots, and in certain cases, particularly with square plots, this may be advantageous. Thus a pattern such as the following might be adopted:

$$
\begin{array}{cccccccccc}
C_1 & + & + & C_2 & + & + & C_3 & + & + & C_4 \\
+ & + & + & + & + & + & + & + & + & + \\
+ & + & + & + & + & + & + & + & + & + \\
C_5 & + & + & C_6 & + & + & C_7 & + & + & C_8
\end{array}
$$

The weighted mean of the four nearest control plots could be taken as the fertility index for plots not in the same row or column as a control, that for the second plot of the second row, for example, being:

$$\tfrac{4}{9}C_1 + \tfrac{2}{9}C_2 + \tfrac{2}{9}C_5 + \tfrac{1}{9}C_6.$$

Clearly many such patterns are possible. Their relative merits can only be tested by extensive examination of uniformity trials on the type of land and with the type of crop on which it is desired to experiment.

IV. Pseudo-factorial arrangements in two equal groups of sets

For two equal groups of sets to be possible the number of varieties must be a perfect square, say p^2. The variety in the u'th set of the first group and the v'th set of the second will be denoted by the pair of numbers uv. Its mean yield over the replicates of the first group of sets will be denoted by x_{uv}, and its mean yield over the replicates of the second group of sets by y_{uv}.

For purposes of computation the mean yields of each group of sets should be set out in squares of the form of Table I.

Table I

First group

Set	1	2	...	p	Mean
	x_{11}	x_{21}	...	x_{p1}	$\bar{x}_{.1}$
	x_{12}	x_{22}	...	x_{p2}	$\bar{x}_{.2}$
	...	...	...	...	...
	x_{1p}	x_{2p}	...	x_{pp}	$\bar{x}_{.p}$
Mean	$\bar{x}_{1.}$	$\bar{x}_{2.}$	...	$\bar{x}_{p.}$	$\bar{x}_{..}$

Second group

Set					Mean
1	y_{11}	y_{21}	...	y_{p1}	$\bar{y}_{.1}$
2	y_{12}	y_{22}	...	y_{p2}	$\bar{y}_{.2}$
...	...	...	...	...	...
p	y_{1p}	y_{2p}	...	y_{pp}	$\bar{y}_{.p}$
Mean	$\bar{y}_{1.}$	$\bar{y}_{2.}$	...	$\bar{y}_{p.}$	$\bar{y}_{..}$

The marginal means, or totals, will be required as indicated. In the actual numerical computations the use of totals is more convenient, but the corresponding algebraic formulae can be more neatly expressed (and remembered) in terms of the means; the substitution of totals in place of means for numerical work will present no difficulty.

The equations for estimating the varietal differences can be derived directly by the classical method of least squares, fitting constants for sets and for varieties. This method will also conveniently provide expressions for the various components of the sum of squares in the analysis of variance. The errors of the various types of comparison are, however, better derived by methods recently developed in connection with the analysis of factorial experiments as described by Fisher (1) and Yates (3). We propose, therefore, to use this line of approach in the present paper.

An experimental arrangement in two equal groups of sets is analogous to a factorial experiment involving two factors (each with p values) in which the main effects of one factor are confounded in one half (the first group) of the replications and the main effects of the other factor are confounded in the other half (the second group). In such an experiment the p^2-1 treatment degrees of freedom are divisible into

First factor	$p-1$
Second factor	$p-1$
Interactions	$(p-1)^2$

The main effects of the first factor will be estimated from the second group of replications only, and those of the second factor from the first group only, the precision of each being one-half that of an unconfounded experiment with the same error variance per plot. The interactions will be unaffected by the confounding and will have estimates of full precision.

In the corresponding pseudo-factorial varietal trial the estimates required are not those of main effects and interactions but differences between single varieties. Such differences must therefore be expressed in terms of the main effects and interactions.

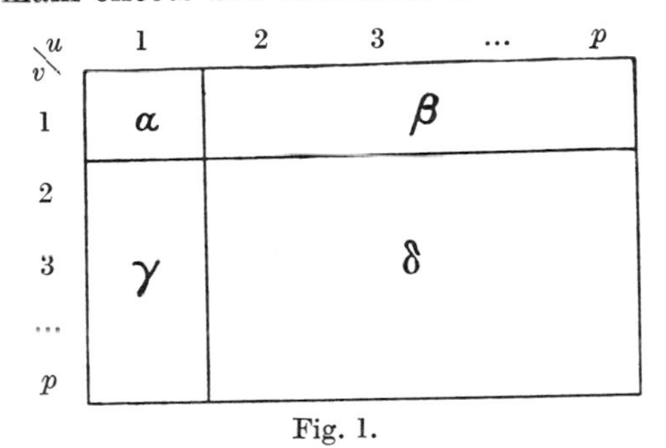

Fig. 1.

We will first consider the estimate of the deviation of any one variety from the mean of all the varieties. Let the true yields per plot of the varieties be represented by τ_{uv}, and take α, β, γ and δ to represent the sums of the τ's contained in the compartments of the two-way table represented in Fig. 1: so that $\alpha=\tau_{11}$, $\beta=\tau_{21}+\tau_{31}+\dots\tau_{p1}$, etc. Further take

$$A=(p-1)^2\,\alpha-(p-1)\,(\beta+\gamma)+\delta,$$
$$B=(p-1)\,(\alpha+\beta)-(\gamma+\delta),$$
$$C=(p-1)\,(\alpha+\gamma)-(\beta+\delta),$$
$$D=\alpha+\beta+\gamma+\delta.$$

The quantity A is a component of the interactions of the two factors and is therefore estimated from both groups of replications, *i.e.* from the mean of the x's and y's. B is a component of the main effects of the second factor and is therefore estimated from the first group of replications, *i.e.* from the x's only. Similarly C is estimated from the second group of replications, *i.e.* from the y's only.

Now $p^2\tau_{11}=A+B+C+D$, and replacing D by $p^2\bar{\tau}_{..}$ we have

$$p^2(\tau_{11}-\bar{\tau}_{..})=A+B+C,$$

and the estimate $t_{11}-\bar{t}_{..}$ of $\tau_{11}-\bar{\tau}_{..}$ is therefore given by

$$\begin{aligned} p^2(t_{11}-\bar{t}_{..})=(p-1)^2\tfrac{1}{2}(\alpha_x+\alpha_y)-(p-1)\tfrac{1}{2}(\beta_x+\beta_y+\gamma_x+\gamma_y)+\tfrac{1}{2}(\delta_x+\delta_y)\\ +(p-1)(\alpha_x+\beta_x)-(\gamma_x+\delta_x)\\ +(p-1)(\alpha_y+\gamma_y)-(\beta_y+\delta_y),\end{aligned}$$

which after some simplification, replacing $\bar{t}_{..}$ by $\frac{1}{2}(\bar{x}_{..}+\bar{y}_{..})$, reduces to the final generalised expression

$$t_{uv}=\tfrac{1}{2}(x_{uv}+y_{uv}-\bar{x}_{u.}+\bar{x}_{.v}+\bar{y}_{u.}-\bar{y}_{.v}).$$

Differences between the adjusted values t_{uv} will give efficient estimates of the varietal differences freed from block effects.

The variance of the difference of two such adjusted means can be determined in a similar manner by considering a 2×2 element of four values.

Consider the four values

$$\begin{matrix}\tau_{11} & \tau_{21}\\ \tau_{12} & \tau_{22}\end{matrix}$$

The differences between these may be represented as the main effects and interactions of this two-way table, say

$$\begin{aligned} U&=\tfrac{1}{2}\{(\tau_{21}+\tau_{22})-(\tau_{11}+\tau_{12})\},\\ V&=\tfrac{1}{2}\{(\tau_{12}+\tau_{22})-(\tau_{11}+\tau_{21})\},\\ \{U.V\}&=\tfrac{1}{2}\{(\tau_{11}+\tau_{22})-(\tau_{12}+\tau_{21})\}.\end{aligned}$$

Now take α', β', γ' and δ' to represent the sums of the τ's contained in the compartments of the two-way table represented in Fig. 2:

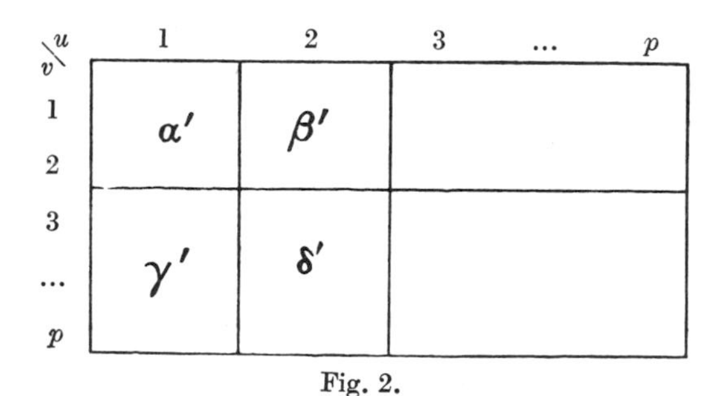

Fig. 2.

so that $\alpha' = \tau_{11} + \tau_{12}$, etc. Further take

$$E = (p-2)(\alpha' - \beta') - 2(\gamma' - \delta'),$$
$$F = \alpha' - \beta' + \gamma' - \delta'.$$

We have

$$-2pU = p(\alpha' - \beta') = E + 2F.$$

The quantity E is a component of interaction between the two factors and therefore its estimate has a variance

$$V(E) = \frac{\sigma^2}{r}\{(p-2)^2 4 + 8(p-2)\} = \frac{4\sigma^2}{r} p(p-2),$$

while F is a component of the main effect of the first factor and therefore its estimate has a variance

$$V(F) = \frac{2\sigma^2}{r}(2p) = \frac{4\sigma^2}{r} p.$$

Hence

$$V(U) = \frac{1}{4p^2}\{V(E) + 4V(F)\} = \frac{\sigma^2}{rp}(p+2) = V(V)$$

and since $\{U.V\}$ is a component of interaction its estimate has a variance of σ^2/r.

Now

$$\tau_{21} - \tau_{11} = U - \{U.V\}$$

and therefore the variance of the difference of two varieties having a set in common is

$$V(t_{21} - t_{11}) = \frac{\sigma^2}{rp}(p+2) + \frac{\sigma^2}{r} = \frac{2\sigma^2}{r} \cdot \frac{p+1}{p},$$

since all the degrees of freedom concerned are orthogonal.

Similarly the variance of the difference of two varieties not having a set in common is

$$V(t_{22} - t_{11}) = V(U+V) = \frac{2\sigma^2}{r} \cdot \frac{p+2}{p}.$$

It can easily be shown that the mean variance of all comparisons is equal to twice the expectation of the sum of squares of deviations between all adjusted varietal means divided by the associated number of degrees of freedom, p^2-1, thus providing a check. Resolving this sum of squares into its component parts we have the following expectations:

	Degrees of freedom	Expectation of mean square	Expectation of sum of squares
First factor	$p-1$	$2\sigma^2/r$	$2\sigma^2(p-1)/r$
Second factor	$p-1$	$2\sigma^2/r$	$2\sigma^2(p-1)/r$
Interactions	$(p-1)^2$	σ^2/r	$\sigma^2(p-1)^2/r$
Total	p^2-1		$\sigma^2(p+3)(p-1)/r$

Thus the mean variance of all comparisons is

$$V_m = \frac{2\sigma^2}{r} \cdot \frac{p+3}{p+1}.$$

Had there been no confounding the variance of every comparison would have been $2\sigma'^2/r$. The factor $(p+3)/(p+1)$ is therefore a measure of the increase in variance that results from the division of the varieties into sets when the error variance per plot is unaltered by the resultant reduction in block size. Its reciprocal, which may be called the *efficiency factor* of the arrangement, is a measure of the inherent strength of the arrangement. Values of this factor for various numbers of varieties are given in Table II, which also includes values of the similar factor for two groups of unequal sets described in section VII, and for the Latin square type of arrangement described in section VI, this latter being

$$(p+1)/(p+2\tfrac{1}{2}).$$

Table II. *Efficiency factors in two-dimensional arrangements*

No. of varieties	25	30	36	42	49	64	100	144	256	400
Arrangement	5^2	5×6	6^2	6×7	7^2	8^2	10^2	12^2	16^2	20^2
Two groups of sets	0·75	0·764	0·778	0·788	0·8	0·818	0·846	0·867	0·895	0·913
Latin square grouping	0·8	—	0·824	—	0·842	0·857	0·880	0·897	0·919	0·933
Random controls	0·485	—	0·511	—	0·532	0·547	0·568	0·596	0·631	0·658

The table also shows the similar efficiency factors for the method of random controls described in section II. Since the error variance per plot is the same for all three types of arrangement the comparison of the three factors gives the relative efficiency of the three methods. It will be seen that the method of random controls is always very much less efficient than the corresponding pseudo-factorial arrangements.

The partition of the degrees of freedom in the analysis of variance is shown in Table III for the case of randomised blocks.

Table III

			Degrees of freedom	
Blocks	Between groups		1	$pr-1$
	Between sets	Group I	$p-1$	
		Group II	$p-1$	
	Within sets	Group I	$p(\frac{1}{2}r-1)$	
		Group II	$p(\frac{1}{2}r-1)$	
Varieties	First factor		$p-1$	p^2-1
	Second factor		$p-1$	
	Interactions		$(p-1)^2$	
Error	Between groups		$(p-1)^2$	$(p-1)(pr-p-1)$
	Within group I		$p(p-1)(\frac{1}{2}r-1)$	
	Within group II		$p(p-1)(\frac{1}{2}r-1)$	
		Total	p^2r-1	

The first factor component of the varietal sum of squares is computed from the values of group II only, and the second factor component from group I only. The varietal sums of squares are therefore equal to

First factor: $\frac{1}{2}r\,(pS\bar{y}_{u.}{}^2 - p^2 S\bar{y}_{..}{}^2)$.

Second factor: $\frac{1}{2}r\,(pS\bar{x}_{.v}{}^2 - p^2 S\bar{x}_{..}{}^2)$.

Interactions: $\frac{1}{4}r\,\{SS\,(x+y)^2 - pS\,(\bar{x}_{u.}+\bar{y}_{u.})^2 - pS\,(\bar{x}_{.v}+\bar{y}_{.v})^2 + p^2\,(\bar{x}_{..}+\bar{y}_{..})^2\}$.

The sum of squares for sets and groups is

$$\tfrac{1}{2}r\,\{p\,(S\bar{x}_{u.}{}^2 + S\bar{y}_{.v}{}^2) - \tfrac{1}{2}p^2\,(\bar{x}_{..}+\bar{y}_{..})^2\}.$$

If these four sums of squares are added together they total

$$\tfrac{1}{4}r\{SS(x+y)^2 + pS(\bar{x}_{u.}-\bar{y}_{u.})^2 + pS(\bar{x}_{.v}-\bar{y}_{.v})^2 - p^2(\bar{x}_{..}-\bar{y}_{..})^2 - p^2(\bar{x}_{..}+\bar{y}_{..})^2\}$$

which is the best form for computation, the sum of squares for varieties being obtained by subtraction of the sum of squares for sets and groups. The sum of squares for error between groups can likewise be obtained by subtraction of this total from the total sum of squares between all x and y, namely

$$\tfrac{1}{2}r\,\{SSx^2 + SSy^2 - p^2\,(\bar{x}_{..}+\bar{y}_{..})^2\}.$$

V. Three-dimensional pseudo-factorial arrangements in three equal groups of sets

The number of varieties must be a perfect cube, say p^3. We shall only consider the type of arrangement in which each variety is specified by three co-ordinate numbers, u, v, w, vw denoting the set out of the p^2 of the first group in which the variety is located, uw and uv the similar sets of the second and third groups respectively. The mean yields of the variety in the $\frac{1}{3}r$ replicates of each of the three groups will be denoted by x_{uvw}, y_{uvw}, z_{uvw} respectively.

Each of these three groups of yields may be arranged in the form of a cube. The marginal means of each of these will be required and can be denoted by an obvious extension of the previous notation. The cube formed from the means, m_{uvw}, of the corresponding values of these three cubes will also be required.

By a similar process to that of the last section it will be found that the adjusted varietal means are given by

$$t_{uvw} = m_{uvw} + \tfrac{1}{2}\,(\bar{m}_{.vw}+\bar{m}_{u.w}+\bar{m}_{uv.}) - \tfrac{1}{2}\,(\bar{m}_{u..}+\bar{m}_{.v.}+\bar{m}_{..w}) - \tfrac{1}{2}\,(\bar{x}_{.vw}+\bar{y}_{u.w}+\bar{z}_{uv.}) + \tfrac{1}{2}\,(\bar{x}_{u..}+\bar{y}_{.v.}+\bar{z}_{..w}),$$

which for computation is best written in the form

$$t_{uvw} = m_{uvw} + c_{.vw} + c_{u.w} + c_{uv.},$$

where $c_{.vw} = \frac{1}{2}\,(\bar{m}_{.vw} - \bar{x}_{.vw} - \bar{m}_{.v.} + \bar{y}_{.v.})$, etc.

Following the same procedure as in the previous section we may represent the differences of the cube of the eight τ's, $\tau_{111} \dots \tau_{222}$, by the main effects U, V, W, and the interactions $\{U.V\}$, $\{U.W\}$, $\{V.W\}$, $\{U.V.W\}$ of the three-way table, each of these quantities being $\frac{1}{4}$ of the sum of four of the τ's less the sum of the other four.

It will be found that

$$V(U)=V(V)=V(W)=\frac{\sigma^2}{2rp^2}(p^2+2p+4),$$

$$V(\{U.V\})=V(\{U.W\})=V(\{V.W\})=\frac{\sigma^2}{2rp}(p+1),$$

$$V(\{U.V.W\})=\frac{\sigma^2}{2r}.$$

Since

$$\tau_{211}-\tau_{111}=U-\{U.V\}-\{U.W\}+\{U.V.W\},$$

$$\tau_{122}-\tau_{111}=V+W-\{U.V\}-\{U.W\},$$

$$\tau_{222}-\tau_{111}=U+V+W+\{U.V.W\},$$

we obtain

$$V(t_{211}-t_{111})=\frac{2\sigma^2}{rp^2}(p^2+p+1),$$

$$V(t_{122}-t_{111})=\frac{\sigma^2}{rp^2}(2p^2+3p+4),$$

$$V(t_{222}-t_{111})=\frac{\sigma^2}{rp^2}(2p^2+3p+6).$$

The mean variance of all comparisons is found to be

$$V_m=\frac{\sigma^2}{r}\,\frac{2p^2+5p+11}{p^2+p+1}.$$

The efficiency of this type of arrangement is therefore

$$2(p^2+p+1)/(2p^2+5p+11).$$

The values of this factor for various numbers of varieties (including some involving unequal sets, as described in section VIII) are shown in Table IV, which also gives the corresponding factors for the method of controls for comparison.

Table IV. *Efficiency factors in three-dimensional arrangements*

No. of varieties Arrangement	64 4^3	80 $4^2 \times 5$	100 4×5^2	125 5^3	216 6^3	343 7^3	512 8^3
Factorial arrangement	0·667	0·684	0·702	0·721	0·761	0·792	0·816
Random controls	0·393	—	—	0·416	0·455	0·484	0·506

The efficiency factors of these three-dimensional arrangements are decidedly lower than those of the two-dimensional arrangements given in Table II. This, and the greater computational labour they entail, must be offset against the greater reduction in block size.

The partition of the degrees of freedom is similar to that already given for the classification into two groups of sets. The sum of squares due to treatments in the analysis of variance can be built up from its components in the same manner as before, but for practical purposes it is best calculated from the formula

$$rSSS\, t_{uvw}\, m_{uvw} - \tfrac{1}{3} rp\, SS\, (\bar{t}_{.vw}\, \bar{x}_{.vw} + \bar{t}_{u.w}\, \bar{y}_{u.w} + \bar{t}_{uv.} \bar{z}_{uv.}).$$

The sum of squares due to the component of error between groups can be calculated by deducting this, and the sums of squares due to sets, groups and mean, which three latter together total

$$\tfrac{1}{3} rp\, SS\, (\bar{x}_{.vw}{}^2 + \bar{y}_{u.w}{}^2 + \bar{z}_{uv.}{}^2),$$

from the total sum of squares arising from the values of the three cubes, namely,

$$\tfrac{1}{3} r\, SSS\, (x_{uvw}{}^2 + y_{uvw}{}^2 + z_{uvw}{}^2).$$

VI. Two-dimensional pseudo-factorial arrangements in three groups of sets forming a Latin square

If a number of varieties forming a perfect square is classified in two ways as in Table I, and in addition in a third way such that every set of the third group contains one and only one variety from each of the sets of the first group and also from each of the sets of the second group, so that the three classifications fulfil the conditions of a Latin square, each set of varieties in each of the three groupings may be compared by means of randomised blocks or a Latin square.

The analytical solution remains comparatively simple. Denote each of the p^2 varieties by three numbers u, v, w (indicating the sets of the three groups to which that variety belongs) and the mean yields for the three groupings by x_{uvw}, y_{uvw} and z_{uvw} respectively, their mean being m_{uvw}. It will be noted that any two of the three numbers are sufficient to specify the variety.

The estimates of the varietal differences are given by differences of the quantities

$$t_{uvw} = m_{uvw} + \tfrac{1}{2}\, (\bar{m}_{u..} + \bar{m}_{.v.} + \bar{m}_{..w}) - \tfrac{1}{2}\, (\bar{x}_{u..} + \bar{y}_{.v.} + \bar{z}_{..w}).$$

The variance of the differences of two varieties occurring in the same set of one of the three groupings is given by

$$V\, (t_{uvw} - t_{uv'w'}) = \frac{2\sigma^2}{r}\left(1 + \frac{1}{p}\right),$$

as in the classification into two groups of sets, and the variance of the difference of two varieties not occurring in the same set is given by

$$V\,(t_{uvw}-t_{u'v'w'})=\frac{2\sigma^2}{r}\left(1+\frac{3}{2p}\right).$$

The mean variance of all comparisons is

$$V_m=\frac{2\sigma^2}{r}\cdot\frac{p+2\frac{1}{2}}{p+1}.$$

The sum of squares for (mean + groups + sets + treatments) is equal to

$$r\left[SSm_{uvw}{}^2+\tfrac{1}{2}p\,\{S\,(\bar{x}_{u..}-\bar{m}_{u..})^2+S\,(\bar{y}_{.v.}-\bar{m}_{.v.})^2+S\,(\bar{z}_{..w}-\bar{m}_{..w})^2\}\right.$$
$$\left.-\frac{p^2}{6}\,\{(\bar{x}_{...}-\bar{m}_{...})^2+(\bar{y}_{...}-\bar{m}_{...})^2+(\bar{z}_{...}-\bar{m}_{...})^2\}\right].$$

This can be divided into its components in the same manner as the classifications into two groups of sets.

It will be noted that the variance of differences between treatments not in the same set is reduced by the comparisons of the third classification, but that the variance of differences between treatments in the same set remains the same, having as before the factor $\left(1+\frac{1}{p}\right)$. Now when the number of varieties is the square of a prime number (and also in certain other cases) further divisions in sets can be made such that each set contains one and only one variety of every set in every other group. (Each group of sets will constitute with each of the other groups a Graeco-Latin square.) If all the $p+1$ such groups of sets are included in the arrangement, every variety occurs with every other variety in one set and one only. Consequently all comparisons are of equal accuracy, and have in fact a variance of $\frac{2\sigma^2}{r}\left(1+\frac{1}{p}\right)$ equal to that of the most accurate comparison in the divisions into two and three groups of sets(4).

VII. Pseudo-factorial arrangements in two unequal groups of sets

So far we have considered only arrangements in which the number of varieties is the same in all sets. This necessitates that the number of varieties included in the experiment is a perfect square or a perfect cube. This limitation may be overcome if the number in each set is different for the different groups. In particular a two-factor classification may have q varieties in each of the p sets of the first group, and p varieties in each of the q sets of the second group, giving pq varieties in all, with u ranging from 1 to p and v from 1 to q.

An additional difficulty immediately presents itself in arrangements of this type. For since the sets in the two groups are of unequal size the blocks or Latin squares by which the comparisons within sets are made will also be of unequal size, and consequently the error variances per plot (σ_x^2 and σ_y^2 say) cannot be assumed equal for both groups.

Take the weights of the varietal means x_{uv} and y_{uv} of the two groups to be λ and μ respectively, so that $\lambda = r/2\sigma_x^2$ and $\mu = r/2\sigma_y^2$. The quantities A, B and C of section IV must be replaced by

$$A = (p-1)(q-1)\alpha - (q-1)\beta - (p-1)\gamma + \delta,$$
$$B = (q-1)(\alpha+\beta) - (\gamma+\delta),$$
$$C = (p-1)(\alpha+\gamma) - (\beta+\delta).$$

In estimates involving both x and y the weighted mean must be taken. Thus the quantity $\frac{1}{2}(\alpha_x + \alpha_y)$ must be replaced by $(\lambda\alpha_x + \mu\alpha_y)/(\lambda+\mu)$, etc. Estimates involving only x or y will remain unchanged.

This leads by the same procedure as before to the expression

$$(\lambda+\mu)\, t_{uv} = \lambda x_{uv} + \mu y_{uv} - \lambda(\bar{x}_{u.} - \bar{y}_{u.}) + \mu(\bar{x}_{.v} - \bar{y}_{.v}).$$

In order to obtain the variance of differences of the t's E must be redefined as

$$E = (q-2)(\alpha' - \beta') - 2(\gamma' - \delta'),$$

giving $-2qU = E + 2F$, and

$$V(E) - \frac{4q(q-2)}{\lambda+\mu}, \quad V(F) = \frac{2q}{\mu};$$

so that

$$V(U) = \frac{1}{q}\left[\frac{2}{\mu} + \frac{q-2}{\lambda+\mu}\right],$$
$$V(V) = \frac{1}{p}\left[\frac{2}{\lambda} + \frac{p-2}{\lambda+\mu}\right],$$
$$V\{U.V\} = \frac{1}{\lambda+\mu}.$$

Thus

$$V(t_{21} - t_{11}) = \frac{2}{q}\left[\frac{1}{\mu} + \frac{q-1}{\lambda+\mu}\right],$$
$$V(t_{12} - t_{11}) = \frac{2}{p}\left[\frac{1}{\lambda} + \frac{p-1}{\lambda+\mu}\right],$$
$$V(t_{22} - t_{11}) = \frac{2}{pq}\left[\frac{q}{\lambda} + \frac{p}{\mu} + \frac{pq-p-q}{\lambda+\mu}\right],$$

the mean variance of all comparisons being

$$V_m = \frac{2}{pq-1}\left[\frac{q-1}{\lambda} + \frac{p-1}{\mu} + \frac{(p-1)(q-1)}{\lambda+\mu}\right].$$

If the difference in weight is ignored and the t's estimated by the unweighted expression of section IV all components estimated from both x and y will be estimated from the unweighted means and will therefore have their variances increased in the ratio

$$\frac{1}{4}\left(\frac{1}{\lambda}+\frac{1}{\mu}\right):\frac{1}{\lambda+\mu}=\frac{(\lambda+\mu)^2}{4\lambda\mu}.$$

Consequently the variances of the differences of the unweighted t's may be immediately obtained by replacing the fraction $1/(\lambda+\mu)$ by $(\lambda+\mu)/4\lambda\mu$ wherever it occurs. The fractional increase of these components will therefore be

$$\frac{(\lambda+\mu)^2}{4\lambda\mu}-1=\frac{(\lambda-\mu)^2}{4\lambda\mu},$$

and the fractional increases in the total variances will therefore always be slightly less than this.

The percentage values of this fraction are

$\lambda:\mu$	2 : 1	3 : 2	4 : 3	5 : 4	6 : 5
Percentage	12·5	4·2	2·1	1·2	0·83

The loss of information by the use of unweighted means is therefore very small provided the error variances of the two groups are not widely different. This is usually the case in practice if p and q are not widely different. Nor is there ever any reason why they should be. With 90 varieties, for instance, a 9×10 arrangement would be chosen in preference to a 6×15 arrangement, since the nearer the arrangement approaches to symmetry the more efficient it will be. With 95 varieties a 10×10 arrangement with 5 additional varieties would undoubtedly be better than a 19×5 arrangement and probably better than a 12×8 arrangement with one additional variety.

When the weights are equal, so that $\lambda=\mu=r/2\sigma^2$, the variances of the varietal comparisons become

$$V\,(t_{21}-t_{11})=\frac{2\sigma^2}{r}\left(1+\frac{1}{q}\right),$$

$$V\,(t_{12}-t_{11})=\frac{2\sigma^2}{r}\left(1+\frac{1}{p}\right),$$

$$V\,(t_{22}-t_{11})=\frac{2\sigma^2}{r}\left(1+\frac{1}{p}+\frac{1}{q}\right),$$

$$V_m=\frac{2\sigma^2}{r}\,\frac{(p+1)\,(q+1)-4}{pq-1}.$$

If the weights are unequal these expressions may be used without appreciable error in place of the exact expressions in all cases in which it is admissible to ignore the differences of weights.

There is another cause of unequal error variance in the two groups. If these occupy different parts of the field it may be that the fertility differences of one part are less adequately eliminated by the blocks than those of the other part, or one part of the field may be more irregular than the other part. Consequently it is advisable to randomise the sets of treatments allotted to the different blocks (in so far as this is compatible with the arrangement of the blocks) whether p and q are unequal or not. This has been done in the examples which follow.

Finally, if the comparisons within sets are made by means of Latin squares instead of randomised blocks, the number of replicates in the two groups will be unequal, being p and q respectively, giving rise to a further inequality in weights, though this inequality is likely to be compensated to a certain extent by a lower error variance in the smaller squares. Even in this case, therefore, weighting will not always be necessary, though it would seem advisable to examine the two components of error within sets before making a decision.

VIII. Three-dimensional pseudo-factorial arrangements in three unequal groups of sets

Let the varieties be classified according to three numbers u, v, w, ranging from 1 to p, 1 to q and 1 to r respectively, the corresponding weights being λ, μ and ν.

Arguments similar to those developed in the last section apply here also. The exact expressions for t_{uvw} are given by

$$(\lambda+\mu+\nu)\, t_{uvw} = \lambda\,(x_{uvw} - \bar{x}_{.vw}) + \frac{\lambda\nu}{\lambda+\mu}\,(\bar{x}_{uv.} - \bar{x}_{.v.}) + \frac{\lambda\mu}{\lambda+\nu}\,(\bar{x}_{u.w} - \bar{x}_{..w})$$

$$+\bar{x}_{u..}\left(\frac{\mu\nu}{\lambda+\mu} + \frac{\mu\nu}{\lambda+\nu}\right) + \text{similar terms in } y \text{ and } z.$$

The variances of the different types of comparison are

$$V\,(t_{211} - t_{111}) = \frac{2}{qr}\left[\frac{1}{\lambda} + \frac{q-1}{\lambda+\mu} + \frac{r-1}{\lambda+\nu} + \frac{(q-1)\,(r-1)}{\lambda+\mu+\nu}\right],$$

$$V\,(t_{122} - t_{111}) = \frac{2}{pqr}\left[\frac{q}{\mu} + \frac{r}{\nu} + \frac{q(p-1)}{\lambda+\mu} + \frac{r(p-1)}{\lambda+\nu} + \frac{qr-q-r}{\mu+\nu} + \frac{(qr-q-r)(p-1)}{\lambda+\mu+\nu}\right],$$

$$V\,(t_{222} - t_{111}) = \frac{2}{pqr}\left[\frac{p}{\lambda} + \frac{q}{\mu} + \frac{r}{\nu} + \frac{qr-q-r}{\mu+\nu} + \frac{pr-p-r}{\lambda+\nu} + \frac{pq-p-q}{\lambda+\mu} + \frac{(p-1)\,(q-1)\,(r-1)+1}{\lambda+\mu+\nu}\right],$$

with similar expressions for $V\,(t_{121} - t_{111})$, etc.

If the differences in weights are ignored the fractional increase in components with divisor $\lambda+\mu$ will be $(\lambda-\mu)^2/4\lambda\mu$ as before, with similar increases with the divisors $\lambda+\nu$ and $\mu+\nu$. The fractional increase in the components with the divisor $\lambda+\mu+\nu$ will be

$$\frac{1}{9}\left(\frac{1}{\lambda}+\frac{1}{\mu}+\frac{1}{\nu}\right)(\lambda+\mu+\nu)-1=\frac{(\mu-\nu)^2}{9\mu\nu}+\frac{(\lambda-\nu)^2}{9\lambda\nu}+\frac{(\lambda-\mu)^2}{9\lambda\mu}.$$

The loss of information due to the use of unweighted means will therefore be of the same order as in the case of two unequal groups of sets. In practice p need not differ by more than one unit from q and r, which can be equal. This makes available a useful set of values for the total number of varieties which is not effectively increased by the inclusion of a greatest difference of 2 units between p, q and r. The possible numbers between 4^3 and 8^3 are:

64, 80, 100, 125, 150, 180, 216, 252, 294, 343, 392, 448, 512.

When the weights are equal the variances of varietal comparisons are:

$$V\,(t_{211}-t_{111})=\frac{\sigma^2/r'}{qr}\,(2qr+q+r+2),$$

$$V\,(t_{122}-t_{111})=\frac{\sigma^2/r'}{pqr}\,(2pqr+pq+pr+qr+2q+2r),$$

$$V\,(t_{222}-t_{111})=\frac{\sigma^2/r'}{pqr}\,(2pqr+pq+pr+qr+2p+2q+2r).$$

$$V_m=\frac{\sigma^2/r'}{pqr-1}\{2\,(p-1)\,(q-1)\,(r-1)+3\,(q-1)\,(r-1)+3\,(p-1)\,(r-1)+3\,(p-1)\,(q-1)+6\,(p-1)+6\,(q-1)+6\,(r-1)\},$$

being the number of replications.

IX. NUMERICAL EXAMPLES

In order to illustrate the various types of arrangement that are discussed in this paper we will take as an example the preliminary results of an experimental orchard of orange trees at the University of California Citrus Experiment Station at Riverside. The orchard was laid out for fertiliser trials, but during the first ten years of its existence was run as a uniformity trial without treatment.

The trial is fully reported and the results discussed by Parker and Batchelor (2). There were 199 plots in all, each of which contained eight trees in a single row, separated by a row of non-experimental oranges and grape fruit from the neighbouring plots. Each tree occupied a square of area 0·011 acre. Every attempt was made to select uniform land and to eliminate variability due to rootstocks and scions.

Table V. *Mean annual yields in pounds per tree*

	M	L	K	J	I	H	G	F	E	D
2	87*	103*	96*	103*	113*				100*	121*
4	95	106	119	104	120				127	140
6	111	110	109	99	140				116	141
8	79	112	99	115	129				139	137
10	98	113	116	110	122				130	128
12	102*	109*	122*	128*	113*				140	133*
14	93	126	137	128	108	132*			133*	138
16	100	116	129	119	113	130			100	107
18	93	115	112	119	120	132			124	105
20	98	116	127	135	125	131			116	105
22	98*	138*	132*	137	138*	116	120*		122	124*
24	103	†	121	119*	115	114*	123		118*	121
26	97	121	104	110	117	115	103			130
28	119	119	129	132	129	138	141			116
30	108	131	148	130	152	153	140			136
32	127	127*	135	134	126	129	133*	152*		139
34	119*	†	140*	134*	121*	139*	135	128		128*
36	135	147	146	147	150	143	127	138		136
38	120	137	119	124	136	142	148	129		133
40	121	138	135	139	142	150	133	131		122
42	114	128	157	146	144	142*	155	153*		138
44	107*	126*	143*	135*	126*	†	144*	165		144*
46						126	155	155		146
48						129	140	140		117
50						134	142	131		115
52						114	150	149*		122
54						114*	136*	†		108*

* Used as controls. † Yield missing.

Table V shows the mean yields of each plot from 1922 to 1927. The arrangement of the values in the table corresponds to the arrangement of the plots on the ground. The values of four of the plots were discarded for various reasons.

Table VI. *Residual mean squares for blocks of various sizes*

No. of plots per block	No. of blocks	Residual mean square	Plots discarded
199	1	247·43	—
64	3	219·15	L 24†, L 34†, H 44†, F 54†, D 2, D 4, D 6
49	4	192·16	F 54†, E 22, E 24
8	24	119·51	As for blocks of 64
7	28	121·65	As for blocks of 49
4	48	100·89	As for blocks of 64
2	96	104·70	As for blocks of 64

† Yield missing.

The residual mean squares after eliminating block differences are shown for blocks of various sizes in Table VI. These particular block sizes are those required for the examples we shall consider.

The residual mean square is reduced consistently by decrease in block size, except for blocks of two plots and of seven plots. About $2\frac{1}{2}$ times as much information is available on comparisons within blocks of four plots as is available on the average of all comparisons.

The mean square for blocks of two plots is not significantly larger than the mean square for blocks of four plots, but the reversal of the tendency to decrease is interesting, and may possibly indicate the existence of competition effects.

The following arrangements will be considered in detail:

(*a*) The comparison of 49 varieties using systematic controls.

(*b*) The comparison of 49 varieties in a 7×7 pseudo-factorial arrangement.

(*c*) The comparison of 64 varieties in a $4 \times 4 \times 4$ pseudo-factorial arrangement.

The relative accuracy of these and other possible arrangements will also be discussed.

(*a*) *Systematic controls*

In order to illustrate the method of systematic controls these have been taken at intervals of five or six plots. The actual plots selected are marked with an asterisk in Table V. The choice was made without reference to the yields, and where necessary by random selection. Missing plots were omitted, though it may be noted that by ill chance all four missing plots would have been chosen as controls.

The weighted means of the two neighbouring control plots may now be calculated for each of the experimental plots. Those of M 4–10, for instance, are 90, 93, 96, 99, and those of H 44–52 are [137·33], 132·67, 128, 123·33, 118·67, that in square brackets corresponding to the missing plot H 44. These means will be taken as the fertility indices of the plots. Fractions may be avoided by the use of multiples in the actual analyses.

The analyses of variance and covariance of the fertility indices and the corresponding actual yields are shown in Table VII, which also shows the analyses of variance for the differences of the indices and the yields and for the quantities $y - bf$. The value of b is given by

$$14817{\cdot}69/19885{\cdot}13 = 0{\cdot}7452.$$

The residual mean square for the differences equals 149·78, and that for the quantities $y-bf$ equals 141·90, the degrees of freedom being reduced by 1 in the second case.

In this example the value of b is not very far removed from unity; the difference between the two errors is consequently not very large, and the covariance method gives only 5 per cent. more information.

Table VII. *Analyses of variance and covariance of fertility indices* (f) *and plot yields* (y)

		Sums of squares and products				
		f^2	fy	y^2	$(y-f)^2$	$(y-bf)^2$
Blocks	2	5271·21	4768·10	4773·04	508·05	593·95
Residual	146	19885·13	14817·69	31617·63	21867·38	20576·02
Total	148	25156·34	19585·79	36390·67	22375·43	21169·97

To make allowance for the loss of information due to the reduction in the number of available plots these error variances must be multiplied by 195/149 (*i.e.* total number of plots/number of non-control plots) before comparison with that obtained in ordinary randomised blocks of 49 plots. The covariance method thus gives a value of 185·70 as compared with 192·16 for randomised blocks of 49 plots.

(*b*) 7 × 7 *pseudo-factorial arrangement*

The plots were divided without reference to the yields into 28 blocks of seven plots each. The particular set of varieties assigned to each block was chosen at random from all the sets for the reasons indicated at the end of section VII. The varieties within each set were then assigned to the plots of the chosen block at random.

Table VIII indicates the resulting arrangement, and should be read in conjunction with Table V. Three of the plots with missing yields have been included in the arrangement, as would be the case in practice.

When some of the yields are missing the correct way of conducting the analysis is to represent these missing yields by symbols, form the error mean square in the ordinary way, and determine the values of the missing yields which make this error mean square a minimum. The analysis can then be completed using these values.

There is no great difficulty in this procedure, but a simpler one is available which will usually give satisfactory results. This consists of giving a value to each missing yield which minimises the part of the error mean square contributed by all the replicates of the set for which the yield is missing. In the example before us there are only two replicates of each set, and this will therefore be effected when the difference between the value of the missing yield and that of the other replicate of the same variety is equal to the mean difference between the two replicates of all other varieties of these two sets.

Table VIII. *Arrangement of blocks and varieties:* 7 × 7 *pseudo-factorial arrangement*

	M	L	K	J	I	H	G	F	E	D
2	77	75	74	73	72				71	76
4	52	73	72	57	44				73	43
6	72	71	42	51	43				63	33
8	32	77	12	55	47				56	23
10	62	75	22	53	45				46	13
12	12	72	62	52	42				26	53
14	22	74	32	54	41	26			16	63
16	42	76	52	56	46	16			76	67
18	23	23	75	71	36	46			36	61
20	13	21	25	31	34	56			66	65
22	73	26	45	61	32	66	55		*	66
24	53	22†	65	11	37	76	51		*	64
26	43	24	15	21	33	36	56			62
28	33	27	35	41	35	11	52			34
30	63	25	55	51	31	17	57			54
32	26	45	14	77	57	15	54	17		64
34	22	46†	64	37	37	13	53	13		74
36	27	42	74	47	47	14	34	14		44
38	21	47	24	67	27	12	32	12		14
40	25	43	34	27	67	16	36	11		24
42	24	44	44	57	17	51	31	16		55
44	23	41	54	17	77	71†	35	15		65
46						61	33	17		75
48						31	37	11		15
50						21	66	13		45
52						41	65	12		35
54						11	64	*†		25

* Discarded. † Yield missing.

Blocks of 49 plots: (1) M, L, K 2–10; (2) K 12–44, J, I 2–20; (3) I 22–44, H, G 22–52; (4) G 54, F 32–52, E, D.

Here the yield of plot 24 of column L is missing. This plot corresponds to variety 22 in the first grouping, and the mean of all the other yields in this block is 123·3. The other replicate of 22 in the first grouping is the plot 34 of column M with yield 119, the mean of all the other plots in the block being 120·7. The difference is $123{\cdot}3-120{\cdot}7=2{\cdot}6$ and the required yield is therefore $119+2{\cdot}6=122$ say. The values of plot 34 of column L and plot 44 of column H are similarly found to be 125 and 118 respectively.

This procedure will slightly inflate the mean squares for treatments and for error between groups, their average value (when treatments produce no effect) and the average value of the mean square for error within groups being approximately in the ratio of 99 : 96. It will, moreover, sacrifice some of the available information on the varieties with missing plots. In this respect it will be roughly equivalent to the process of estimating a quantity ζ from three measures z_1, z_2 and z_3 of equal

Table IX. 7 × 7 *arrangement: sums of yields (less* 100) *of pairs of replicates in same grouping*

First group ($2x_{uv}$)

Set	1	2	3	4	5	6	7	Total
	69	36	107	34	22	45	10	323
	71	41*	86	60	69	79	22	428
	...	...	...	...	...	...	...	...
Total	590	305	498	373	298	433	115	2612

Second group ($2y_{uv}$)

Set								Total
1	33	44	64	46	72	63	37*	359
2	1	9	16	9	24	20	30	109
	...	...	...	...	...	...	...	...
Total	291	306	336	391	439	349	298	2410

* Including estimate of missing value.

Sum of both groups

$v \backslash u$	1	2	3	4	5	6	7	Total	$\frac{1}{2}(\bar{x}_{.v}-\bar{y}_{.v})$
1	102	80	171	80	94	108	47	682	− 1·3
2	72	50	102	69	93	99	52	537	+11·4
	...	...	...	...	...	...	...	...	...
Total	881	611	834	764	737	782	413	5022	+ 7·2
$\frac{1}{2}(\bar{y}_{u.}-\bar{x}_{u.})$	−10·7	0·0	−5·8	+0·6	+5·0	−3·0	+6·5	−7·2	

Adjusted values of treatment means (t_{uv})

$v \backslash u$	1	2	3	4	5	6	7
1	113·5	118·7	135·6	119·3	127·2	122·7	117·0
2	118·7	123·9	131·1	129·2	139·6	133·2	130·9
	...	...	...	...	...	...	...

accuracy by the expression $\frac{1}{2}\{\frac{1}{2}(z_1+z_2)+z_3\}$, instead of the mean $\frac{1}{3}(z_1+z_2+z_3)$. The variances of these two expressions being in the ratio $\frac{3}{8}:\frac{1}{3}$ or $\frac{9}{8}$, the amount of information lost will be $\frac{1}{9}$, which is not likely to be serious unless the variety involved turns out to be of special interest. If there are six replications, instead of four, the corresponding ratio is $\frac{5}{24}:\frac{1}{5}$ or $\frac{25}{24}$.

Table IX shows the sum of the two replicates of each variety for each grouping, arranged in the same manner as Table I. In order to save space only the values for varieties having $v=1$ or 2 are included. The sum of these two tables is also shown, together with the quantities $\frac{1}{2}(\bar{y}_{u.}-\bar{x}_{u.})$ and $\frac{1}{2}(\bar{x}_{.v}-\bar{y}_{.v})$, the first of the former of these quantities, $-10{\cdot}7$, for example, being equal to $\frac{1}{28}(291-590)$. From these values the adjusted values of the varietal means are derived. These are shown in the last line of Table IX, that for variety 11, for example, being given by

$$\tfrac{1}{4}.102-10{\cdot}7-1{\cdot}3+100=113{\cdot}5.$$

The analysis of variance is shown in Table X. The sum of squares for (mean + groups + sets + varieties) is equal to

$$142{,}788+6299{\cdot}86+5370{\cdot}71-208{\cdot}18=154{,}250{\cdot}39,$$

the terms being given in the same order as in the formula of section IV. The sum of squares for (mean + groups + sets) is 147,149·71. The difference gives the sum of squares for varieties, and the difference of the first quantity from the total sum of squares arising from all x_{uv} and y_{uv}, namely 158,077, gives the error between groups.

Table X. *Analysis of variance: 7 × 7 arrangement*

	Degrees of freedom	Sum of squares	Mean square	
Blocks (including groups and sets)	27	27928·63	1034·39	
Varieties	48	7100·68	147·93	
Error: Between groups	36	3826·61	106·29	121·64
Within groups	81*	9144·15	112·89	
Total	192*	48000·07	250·00	

* 3 deducted for missing yields.

The estimates of error between and within groups are about equal, as they should be. The combined error mean square, 110·86, may be compared with the treatment mean square, 147·93, by means of the z test. The value of z obtained is 0·1440, whereas the 5 per cent. point for these numbers of degrees of freedom is equal to 0·1913. The observed value (which is slightly inflated because of the method of estimating the missing plots) is thus well within the range that may be expected by chance.

The standard errors of the differences of the adjusted varietal means are as follows:

Two varieties in same set (row or column):

$$\sqrt{\frac{2\times 110{\cdot}86}{4}\left(1+\tfrac{1}{7}\right)}=\pm 7{\cdot}96.$$

Two varieties not in same set:

$$\sqrt{\frac{2\times 110{\cdot}86}{4}\left(1+\tfrac{2}{7}\right)}=\pm 8{\cdot}44.$$

If one of the varieties has a missing yield the standard errors will be increased approximately in the square root of the ratio $(\frac{1}{4}+\frac{3}{8}):(\frac{1}{4}+\frac{1}{4})$ or $\sqrt{5/4}$, and if both the varieties have missing yields by $\sqrt{3/2}$.

In order to compare the accuracy of this pseudo-factorial arrangement with an arrangement in randomised blocks the pooled estimate of error from varieties and error should be taken. This is 121·64. The corresponding error for randomised blocks of 49 plots is 192·16. The ratio $192{\cdot}16 : 121{\cdot}64 = 1{\cdot}580$, multiplied by the factor $(p+1)/(p+3)$ or 8/10, gives 1·264, so that the pseudo-factorial arrangement is 26 per cent. more efficient. If a 7×7 Latin-square pseudo-factorial arrangement had been adopted (this would require a multiple of 3 replications) the factor would be $(p+1)/(p+2\frac{1}{2})$ or 16/19, giving the ratio 1·331, *i.e.* a gain of a further 7 per cent.

Table XI. *Arrangement of blocks and varieties: $4\times 4\times 4$ pseudo-factorial arrangement*

	M	L	K	J	I	H	G	F	E	D
2	342	421	424	234	134				142	*
4	242	423	422	334	434				144	*
6	142	341	442	231	223				143	*
8	442	311	432	431	224				141	211
10	213	321	422	331	221				243	213
12	313	331	412	131	222				143	214
14	413	112	222	124	143	233			343	212
16	113	132	232	114	133	243			443	432
18	323	122	242	144	113	213			111	132
20	333	142	212	134	123	223			113	232
22	343	244	441	114	131	414	342		114	332
24	313	*†	421	214	141	411	341		112	222
26	431	224	431	414	121	412	344			122
28	434	214	411	314	111	413	343			322
30	432	234	424	444	321	121	313			422
32	433	241	444	443	323	221	312	433		322
34	323	*†	434	441	324	421	314	423		312
36	423	231	414	442	322	321	311	413		332
38	223	211	244	324	433	121	224	443		342
40	123	221	242	344	233	124	424	334		232
42	412	312	241	334	133	122	124	333		231
44	212	112	243	314	333	*†	324	332		233
46						123	132	331		234
48						141	134	411		144
50						341	133	111		444
52						241	131	211		344
54						441	311	*†		244

* Discarded. † Yield missing.
Blocks of 64 plots: (1) M, L, K; (2) J, I, H; (3) G, F, E, D.

(c) $4\times4\times4$ pseudo-factorial arrangement

For this arrangement the plots were divided without reference to the yields into 48 blocks of four plots each, and the 64 varieties assigned at random as in the 7×7 arrangement. Missing plots were, however, excluded. The actual arrangement obtained is shown in Table XI.

The first step in the computations is to set out the yields in order as in Table XII, where the yields of the first group and the sum of all the groupings are shown for varieties having $v=1$ or 2, together with their marginal totals. The yields of the other two groupings and of varieties having $v=3$ or 4, have been omitted to save space.

The next step is to prepare a table of the quantities $c_{.vw}$, $c_{u.w}$ and $c_{uv.}$. These are also shown in Table XII. Thus

$$-3{\cdot}38=\tfrac{1}{24}\,365-\tfrac{1}{8}156-\tfrac{1}{96}1178+\tfrac{1}{32}322.$$

Finally a table of the adjusted yields may be prepared by adding these corrections to one-third of the sum of all the groupings. Thus

$$132{\cdot}16=100+\tfrac{1}{3}(84)-3{\cdot}38+6{\cdot}03+1{\cdot}51.$$

The analysis of variance is shown in Table XIII. The sum of squares due to treatments is given by the sum of the products of the adjusted values and the variety totals of Table XII, less the sum of the products of the block totals and the corresponding marginal means of the adjusted values, $\bar{t}_{.vw}$, etc. (not shown). Omitting the 100's from the t's the first sum of products is 130,560·81, while the second is given by the sum of $27{\cdot}98\times156$, etc., and two similar sums for y and z, which three together total 124,189·42. The computation of the other sums of squares requires no comment.

The standard error of the difference of two varieties occurring in the same set (*i.e.* having two of the three numbers u, v, w the same) is

$$\sqrt{\frac{2\times100{\cdot}69\times21}{48}}=\pm9{\cdot}39,$$

that of two varieties having one of the three numbers u, v, w the same is

$$\sqrt{\frac{100{\cdot}69\times48}{48}}=\pm10{\cdot}03,$$

and that of two varieties having no common number is

$$\sqrt{\frac{100{\cdot}69\times50}{48}}=\pm10{\cdot}24.$$

To compare the accuracy of the $4\times4\times4$ arrangement with an arrangement in ordinary randomised blocks of 64 plots the pooled residual

Table XII. $4 \times 4 \times 4$ *arrangement*

w	1					2					3					4					Sum				
u/v	1	2	3	4	$.vw$	1	2	3	4	$.vw$	1	2	3	4	$.vw$	1	2	3	4	$.vw$	$uv.$				$.v.$
Yields of first group (x_{uvw})																									
1	31	49	36	40	156	26	7	28	14	75	0	−2	2	−7	−7	37	19	32	10	98	94	73	98	57	322
2	53	29	43	39	164	30	21	16	36	103	21	20	19	35	95	55	48	44	33	180	159	118	122	143	542
	...	...	...	...	...	...	...	...	...	...	...	...	...	...	...	...	...	...	...	...	...	...	...	...	...
	141	91	123	108	463	72	28	55	36	191	105	90	80	64	339	122	78	102	78	380	440	287	360	286	1373
	$u.w$				$..w$	$u.w$				$..w$	$u.w$				$..w$	$u.w$				$..w$	$u..$				...
Sum of three groups ($3m_{vuw}$)																									
1	84	123	75	83	365	70	72	89	51	282	36	58	45	69	208	78	71	102	72	323	268	324	311	275	1178
2	112	89	108	63	372	87	71	105	71	334	72	91	38	69	270	133	98	89	77	397	404	349	340	280	1373
	...	...	...	...	...	...	...	...	...	...	...	...	...	...	...	...	...	...	...	...	...	...	...	...	...
	395	394	324	242	1355	260	241	359	171	1031	271	370	232	316	1189	362	314	336	308	1320	1288	1319	1251	1037	4895
Corrections ($c_{.vw}$, $c_{u.w}$, $c_{uv.}$)																									
					−3·38					+3·29					+10·45					+2·12	+1·51	−8·27	−5·70	−0·78	
					−6·98					−0·94					−2·61					−7·94	−2·83	−3·23	−6·24	+6·80	
					...					...					...					...	...	...	...	...	
	+6·03	−0·26	+9·95	+1·28		+5·97	+1·18	+2·22	+5·64		+6·91	+3·66	+14·53	−1·34		+1·74	−1·26	−4·72	−9·01						

Adjusted treatment means (t_{uvw})

132·16	129·09	125·87	124·79	134·10	120·20	129·48	125·15	130·87	125·17	134·28	131·33	131·37	116·26	125·70	116·33
133·55	119·20	132·73	122·10	131·20	120·68	130·04	135·17	125·47	128·15	118·35	125·85	135·30	120·24	110·77	115·52
...	...	...	...	...	...	...	...	...	...	...	...	...	...	...	...

variance of varieties and error, 100·89, must be multiplied by the factor $\frac{63}{42}$, giving 151·34. The ratio of the residual variance of blocks of 64 plots, 219·15, to this is 1·448, so that the pseudo-factorial arrangement is 45 per cent. more efficient.

Table XIII. *Analysis of variance:* $4 \times 4 \times 4$ *arrangement*

	Degrees of freedom	Sum of squares	Mean square	
Blocks (groups and sets)	47	33004·74	702·23	
Varieties	63	6371·39	101·13	100·89
Error between groups	81	8155·86	100·69	
Total	191	47531·99	248·86	

If instead of a $4 \times 4 \times 4$ arrangement an 8×8 arrangement had been adopted (an even number of replications would be required for this), the residual variance of blocks of eight plots, 119·51, would have to be multiplied by the factor $\frac{11}{9}$, giving a ratio of 1·500. With an 8×8 Latin-square pseudo-factorial arrangement the factor would be $\frac{21}{18}$, giving a ratio of 1·572.

X. Comparison of the relative efficiencies of the designs of section IX

A summary of the efficiencies of the various arrangements discussed in the last section, relative to that of ordinary randomised blocks, is given in Table XIV. The efficiencies in the limiting case in which the error variance is the same for all block sizes, *i.e.* when there is no association between neighbouring plots, are also given for comparison.

The pseudo-factorial arrangements have proved decidedly the most efficient in every case, more especially with 64 varieties. Even the $4 \times 4 \times 4$ arrangement, which was really only included to illustrate the computations, has been very successful, though for only 64 varieties and a small number of replications the 8×8 arrangements are to be preferred. Three-dimensional arrangements are likely to be most advantageous with a larger number of varieties, and in cases in which there are sufficient replications for the randomised blocks to be replaced by Latin squares.

It is not claimed that the uniformity trial on which the examples are based is necessarily typical of the soil heterogeneity ordinarily met with. It was in fact the first uniformity trial encountered in a search of the literature which contained the requisite number of plots of reasonable size. A certain number of trials were inspected which contained a large number of small plots ($\frac{1}{500}-\frac{1}{200}$ acre), and none of these exhibited such a large degree of soil heterogeneity. It may be doubted, however, whether trials of this nature fairly represent the type of land ordinarily available for experiment: in some cases, at least, it would seem that a specially uniform piece of land was chosen, often after inspection of the growing crop.

In any case the values of Table XIV for the case where there is no association between plots show that the pseudo-factorial arrangements

Table XIV. *Comparison of various types of arrangement.* (*Efficiencies expressed as a percentage of that of ordinary randomised blocks*)

	Chosen example	No association between plots
A. 49 varieties		
Randomised blocks of 49 plots	100	100
Blocks of 7 plots with 2 controls per block	84·0	53·2
Systematic controls every 5th or 6th plot (differences)	98·0	47·7
Systematic controls every 5th or 6th plot (covariance)	103·5	76·4
7 × 7 pseudo-factorial arrangement	126·4	80
7 × 7 Latin-square pseudo-factorial arrangement	133·1	84·2
B. 64 varieties		
Randomised blocks of 64 plots	100	100
Blocks of 8 plots with 2 controls per block	100·3	54·7
Systematic controls every 5th or 6th plot (differences)	111·9	47·7
Systematic controls every 5th or 6th plot (covariance)	118·0	76·4
4 × 4 × 4 pseudo-factorial arrangement	141·8	66·7
8 × 8 pseudo-factorial arrangement	150·0	81·8
8 × 8 Latin-square pseudo-factorial arrangement	157·2	85·7

are never likely to be much less efficient than systematic controls, and are always decidedly more efficient than random controls. They may be somewhat less efficient than ordinary randomised blocks when there is little soil heterogeneity to eliminate.

XI. Summary

A new method of arranging variety trials involving large numbers of varieties is described. This type of arrangement, for which the name *pseudo-factorial* arrangement is proposed, enables the block size to be kept small without the use of controls.

Various possible types of pseudo-factorial arrangement are discussed in detail and the necessary formulae developed. The appropriate methods of computation are illustrated by numerical examples based on the results of a uniformity trial on orange trees. It is shown that pseudo-factorial arrangements are likely to be more efficient than arrangements involving the use of controls. In cases where there is considerable soil heterogeneity they are also markedly more efficient than randomised blocks containing all the varieties. In the chosen example gains in efficiency ranging from 26 to 57 per cent. were obtained.

REFERENCES

(1) Fisher, R. A. *The Design of Experiments* (1935). Edinburgh: Oliver and Boyd.
(2) Parker, E. R. and Batchelor, L. D. *Hilgardia* (1932), **7**, No. 2.
(3) Yates, F. *Supp. J. R. Statist. Soc.* (1935), **2**, 181.
(4) —— *Ann. Eugen.* (1936), **7** (in the Press).

Paper VII

INCOMPLETE RANDOMIZED BLOCKS

From ANNALS OF EUGENICS
Volume VII, Part II, pp. 121–140, 1936

Author's Note

Paper VII develops the idea, mentioned in Paper V, of balanced incomplete blocks. It is in a sense complementary to Paper VI, but differs from it in that there is not necessarily any factorial structure. Instead the restriction is imposed that blocks must contain each pair of treatments equally frequently. As might be expected such designs have a very simple least square solution and all treatment comparisons are of equal accuracy.

The problem of finding designs which satisfy the required conditions presents interesting combinatorial problems which have since been extensively explored by other workers. Tables of known designs have been published in successive editions of Fisher and Yates. A further extension only briefly hinted at in this paper is that of partially balanced designs, which by somewhat relaxing the above restriction permit greater flexibility in block size and number of treatments.

1969 FRANK YATES

INCOMPLETE RANDOMIZED BLOCKS

By F. YATES, M.A.
Rothamsted Experimental Station

Introduction

Most biological workers are probably by now familiar with the methods of experimental design known as randomized blocks and the Latin square. These were originally developed by Prof. R. A. Fisher, when Chief Statistician at Rothamsted Experimental Station, for use in agricultural field trials. The same principles of design are of general utility, wherever the basic material is at all variable, for comparing the effects of different treatments, or the growth of different strains of an organism, or other similar problems.

In a randomized block arrangement each treatment occurs equally frequently, usually once, in each block, the treatments being assigned at random to the experimental units within the block. The word block may be extended to denote any group containing the requisite number of experimental units, and by arranging the grouping so that similar experimental units occur in the same block the accuracy of the treatment comparisons is considerably enhanced, since differences due to dissimilarities between the different blocks are eliminated from these comparisons. The process of random arrangement within the blocks ensures that no treatment shall be unduly favoured, and, moreover, enables an unbiased estimate of experimental error to be obtained, which is itself the basis of valid tests of significance.

A Latin square arrangement is similar in principle to a randomized block arrangement, but in a Latin square two cross-groupings of the experimental units are made, corresponding to the rows and columns of a square, and the treatments are so arranged that each occurs once and once only in each row and in each column. Thus differences between rows and columns (which may represent any desired groupings) are eliminated from the treatment comparisons. The appropriate randomization process consists of taking any square arrangement which satisfies the conditions of a Latin square and rearranging either the rows or the columns, or both, at random, and then allotting the treatments at random.

In agricultural field trials the elimination of differences of soil fertility is of great importance, and for this purpose randomized block and Latin square arrangements have proved eminently suitable. In such experiments there is no definite natural limit to the number of experimental units, here plots, which may be included in a block, or in a row or column of a Latin square. There is thus no definite limit to the number of treatments which may be comprised by an experiment, though naturally the effectiveness of the blocks or rows and columns of a square in eliminating fertility differences becomes less as the number of treatments increases.

In some other experimental material the groupings which most effectively eliminate heterogeneity are more definitely limited in numbers. Pigs cannot be relied on to produce more than six or eight suitable young in a litter. Monozygotic twins form a natural experimental grouping in man. In the local lesion method of studying the virulence of different preparations of virus by inoculating the leaves of susceptible plants and counting the number of lesions produced there are only from two to five suitable leaves to a plant (depending on the species), and only two treatments can be applied to each leaf, one to each half.

In factorial experiments, that is in experiments with all combinations of two or more sets of treatment factors, a method, known as confounding, has been devised whereby the block size may be kept moderate. In confounded arrangements the treatments of each replication are allotted to two or more sub-blocks in such a manner that some or all of the information is sacrificed on unimportant comparisons, usually high-order interactions between the different factors.

The present paper describes another possible modification of the randomized block type of arrangement, analogous to confounding, which enables us to dispense with the restriction that the number of experimental units in a block shall equal the number of treatments. In this modified type of arrangement the number of experimental units per block is fixed, being less than the number of treatments, and the treatments are so allotted to the blocks that every two treatments occur together in a block equally frequently. With six treatments a, b, c, d, e, f, and blocks of three experimental units, for example, the following grouping of treatments into ten blocks satisfies the above conditions:

1	a	b	c	6	b	c	f
2	a	b	d	7	b	d	e
3	a	c	e	8	b	e	f
4	a	d	f	9	c	d	e
5	a	e	f	10	c	d	f

In this set of groupings every two treatments occur together in a block twice, there being five replications of each treatment.

It is proposed to call this type of arrangement a *symmetrical incomplete randomized block arrangement*, or more briefly, when the symmetry and randomization are understood, an *incomplete block arrangement*. Although groupings of the experimental treatments fulfilling the required condition are always possible, whatever the number of treatments and of units in a block, the number of groupings (and therefore of blocks and units) required is likely to be large where the number of treatments is at all large. In a later section we shall investigate the smallest number of groupings required.

A further advantage of symmetrical incomplete block arrangements, which may be of importance in certain lines of research, is that the differences between the various blocks can be properly estimated. Thus if the blocks represent litters, differences between different litters can be studied simultaneously with the investigation of the effects of treatments on the individual animals. If the experiment were arranged in ordinary randomized blocks including animals from more than one litter no proper estimate of differences between litters would be readily available: the use of the residuals after eliminating treatment effects always tends to reduce the apparent differences between litters.

The simplest and most important example of a symmetrical incomplete randomized block arrangement is that in which each block contains only two units, and all possible combinations of the treatments, taken in pairs, are represented in the different blocks. Thus with six treatments there are $\frac{1}{2}(6 \times 5)$ or 15 pairs, and the number of blocks must therefore be some multiple of 15. It will be shown that such an arrangement is decidedly more efficient than the customary method of arrangement in such cases, namely that of treating one treatment as a control and comparing each other treatment with it. In the proposed type of arrangement all treatment comparisons are of the same accuracy, and it is a remarkable fact that this accuracy is the same as that obtained on the comparisons between the control and the other treatments in an arrangement of the customary type using the same number of experimental units.

An experimental arrangement is not likely to be of much use in practice if it involves long and tedious calculations to analyse the results. Fortunately in the case of symmetrical incomplete blocks the procedure of the analysis of variance is very little more complicated than in the case of ordinary randomized block arrangements. The formulae required in the analysis of the results are given in the following sections. They may all be derived by a straightforward application of the method of least squares.

The chief disadvantage of symmetrical incomplete blocks is, as mentioned above, that the number of replications required is in most cases large when the number of treatments is at all large. This disadvantage can be overcome by dispensing with the condition of symmetry, but at the cost of some small loss of efficiency and the inconvenience of having slight variations in accuracy for different sets of treatment comparisons. The simplest arrangement of this type is that in which the treatments (p^2 in number) are set out in the form of a square and two sets of p blocks are formed from the rows and the columns respectively of this square. With pq treatments the square may be replaced by a rectangle. Such arrangements, and analogous arrangements derived from a cube, are eminently suitable for variety trials involving large numbers of varieties. A description is given in (5).

A further analogous type of arrangement is available in the case of a Latin square. If in a Latin square arrangement one row, one column, or one treatment is missing, or one row and one column, or one row or column and one treatment, not only is the analysis of the results quite simple, but the estimate of error is unbiased, so that such arrangements give perfectly valid tests of significance. The analytical procedure is of considerable interest to agriculturalists, who are occasionally confronted with Latin square arrangements in which a row, column or treatment has failed, and a full account of this procedure has therefore been given in (4). The worker who wishes to lay down an incomplete Latin square, or to analyse one, is referred to that paper. No simple analysis appears to exist when two or more rows, columns or treatments are missing, nor when a single row, column and treatment are all missing.

Symmetrical pairs

The case in which the blocks contain two experimental units only is capable of specially simple treatment. In this case all combinations of the treatments two at a time are required for symmetry, so that if there are t treatments there must be some multiple of $\frac{1}{2}t\,(t-1)$ blocks. With r replicates of each treatment each *pair* of treatments is replicated $r/(t-1)$ or say r' times.

If the treatments are a, b, c, etc., and $S\,[a-b]$ represents the total of the r' differences of the treatments a and b in the blocks in which they occur together, we may set out these totals as in Table I.

Table I

	—	$S\,[b-a]$	$S\,[c-a]$	$S\,[d-a]$	...
	$S\,[a-b]$	—	$S\,[c-b]$	$S\,[d-b]$	...
	$S\,[a-c]$	$S\,[b-c]$	—	$S\,[d-c]$	...
	$S\,[a-d]$	$S\,[b-d]$	$S\,[c-d]$	—	...
	...	...	...	...	...
Column totals	T'_a	T'_b	T'_c	T'_d	

In this table each difference total is repeated twice, with opposite signs.

The differences of the quantities

$$\frac{1}{r't}T'_a,\quad \frac{1}{r't}T'_b,\quad \ldots,$$

represent the treatment differences, in units of the effect on a single experimental unit. Since the sum of these quantities is zero they may conveniently be increased by the mean of the experimental values when presenting the final results.

If σ_d^2 is the variance of the *difference* of two experimental values from the same block the standard error of the difference of any two of the above quantities is

$$\sqrt{2} \times \sigma_d \sqrt{\frac{1}{r't}}.$$

If one of the treatments had been used as a control and compared with each of the other treatments in turn there would have been $rt/2\,(t-1)$ replicates of each pair. The standard error of the difference of the control and any other treatment would then be $\sigma_d \sqrt{2\,(t-1)/rt}$ or $\sigma_d \sqrt{2/r't}$, as above, and the standard error of all other comparisons would be $\sqrt{2}$ times this. When the number of treatments is large, therefore, the symmetrical arrangement is almost twice as efficient as the arrangement using one treatment as a control.

If the anlysis of variance is performed in units of a single *difference* between two experimental units, giving an estimate of σ_d^2 directly, the sum of squares due to treatments is

$$\frac{1}{r't}\,(T_a'^2 + T_b'^2 + \ldots),$$

and the total error sum of squares, with $\frac{1}{2}\,(t-1)\,(r't-2)$ degrees of freedom, is the difference of this and the total sum of squares of all differences (without any correction for the mean).

Example of symmetrical pairs

In an experiment to determine the degree of variability in the number of lesions produced when leaves of young plants of *Nicotiana glutinosa* were inoculated with a suspension containing tobacco mosaic virus, fifty plants were taken, each with five leaves, and the numbers of lesions produced on each half-leaf were counted. The numbers observed on the leaves of the first six plants are shown in Table II. Considerable association between halves of the same leaf is apparent.

Table II. *Number of lesions on half-leaves*

Plant		Leaf 1	Leaf 2	Leaf 3	Leaf 4	Leaf 5
1	*L*	*e* 26	*d* 16	*c* 21	*b* 11	*e* 12
	R	*b* 40	*b* 26	*e* 14	*c* 16	*a* 12
2	*L*	*b* 34	*c* 69	*c* 42	*a* 22	*e* 19
	R	*a* 49	*a* 68	*d* 35	*d* 31	*d* 25
3	*L*	*b* 28	*c* 83	*a* 24	*c* 12	*c* 9
	R	*e* 11	*d* 58	*e* 27	*a* 15	*b* 13
4	*L*	*b* 12	*a* 23	*b* 28	*e* 15	*d* 5
	R	*a* 26	*d* 20	*d* 34	*c* 13	*e* 11
5	*L*	*a* 14	*b* 17	*b* 5	*b* 11	*d* 4
	R	*e* 16	*d* 12	*e* 10	*c* 8	*e* 6
6	*L*	*a* 5	*d* 17	*c* 16	*a* 12	*a* 7
	R	*c* 15	*c* 15	*e* 18	*d* 15	*b* 10

An arrangement to compare five suspensions by the method of symmetrical pairs has been imposed on these values, the allocation of pairs of treatments being at random.* The

**Author's addition (1969)* — . . . random within pairs of plants.

actual arrangement arrived at is shown in the table, the treatments being denoted by $a \dots e$.

The possibility in experiments of this kind of using different dilutions of all suspensions to be compared should be borne in mind, but need not be discussed here. The best function of the numbers of lesions to use in the analysis also requires consideration. Here we have taken differences of the square roots.

These differences are shown in Table III, together with their totals. Table IV shows the calculation of the treatment effects, the estimated differences in the square roots per half-leaf being given by the differences of the quantities in the last line.

Table V shows the analysis of variance. The treatment sum of squares is given by the sum of the squares of the totals of Table IV divided by 15. The sums of squares between different pairs and between replicates of pairs are obtained from Table III in the ordinary manner. The component of error from Table IV is obtained by subtraction of the treatment sum of squares from the sum of squares between different pairs. There are no significant

Table III. *Differences of square roots of numbers of lesions*

Plants	$a-b$	$a-c$	$a-d$	$a-e$	$b-c$	$b-d$	$b-e$	$c-d$	$c-e$	$d-e$
1, 2	+1·2	−0·1	−0·9	0·0	−0·7	+1·1	+1·2	+0·6	+0·9	+0·6
3, 4	+1·6	+0·4	+0·3	−0·3	+0·6	−0·5	+2·0	+1·5	−0·3	−1·1
5, 6	−0·6	−1·7	−0·4	−0·3	+0·5	+0·6	−1·0	−0·2	−0·2	−0·4
Total	+2·2	−1·4	−1·0	−0·6	+0·4	+1·2	+2·2	+1·9	+0·4	−0·9

Table IV. *Estimation of treatment effects*

	a	b	c	d	e
	—	−2·2	+1·4	+1·0	+0·6
	+2·2	—	−0·4	−1·2	−2·2
	−1·4	+0·4	—	−1·9	−0·4
	−1·0	+1·2	+1·9	—	+0·9
	−0·6	+2·2	+0·4	−0·9	—
Total	−0·8	+1·6	+3·3	−3·0	−1·1
Total/15	−0·053	+0·107	+0·220	−0·200	−0·073

Table V. *Analysis of variance*

	Degrees of freedom	Sum of squares	Mean square
Between different pairs:			
Total	10	6·3933	0·6393
Treatments	4	1·6200	0·4050
Remainder	6	4·7733	0·7956
Between replicates	20	16·9467	0·8473
Total error	26	21·7200	0·8354
Total	30	23·3400	0·7780

differences between the mean squares, as is to be expected, since the treatments are dummy. The estimated standard error of the differences of the quantities in the last line of Table IV is $\sqrt{2 \times 0{\cdot}8354/15}$ or $\pm 0{\cdot}334$.

In an example of this kind we are not normally interested in the comparisons of the numbers of lesions on different leaves, but if such a comparison is to be made the effects of treatments must clearly be allowed for. This can be done very simply by subtracting from the square roots of the actual counts the values given in the last line of Table IV for the appropriate treatments. Thus the first leaf of the first plant has treatments e and b, the sum of the square roots of the half-leaf counts is 11·4, and the adjusted value of this sum is therefore

$$11{\cdot}4-0{\cdot}107+0{\cdot}073=11{\cdot}366,$$

a practically identical value, as it should be since the treatments are without effect.

If the treatments are arranged in a 5×5 Latin square on the whole leaves of the first five plants, so that leaf positions as well as plants are equalized for each treatment, the accuracy will be very considerably less. The residual mean square of the square roots of the number of lesions on whole leaves after eliminating plants and leaf positions is found to be 1·6116. Dividing this by 5 will give the mean square error for the mean of five replicates, and dividing by a further 2 (instead of 4, since square roots of lesions on whole leaves have been taken) and multiplying by 5/6 to allow for the smaller number of plants used gives a mean square error of 0·1343, which is comparable with the mean square error 0·7780/15 or 0·0519 derived from the total mean square of Table V. The method of symmetrical pairs therefore gives 2·59 times the information given by the Latin square arrangement.

There are types of arrangement other than symmetrical pairs which make use of the association between halves of the same leaf, and which are consequently more efficient than the Latin square arrangement on whole leaves, but there is no need to discuss these here. A general discussion of the relative efficiency of symmetrical incomplete randomized blocks is given later in the paper.

General case: formulae

The following symbols will be used:

Number of treatments: t.

Number of experimental units per block: k.

Number of replications of each treatment: r.

Number of blocks: b.

Total number of experimental units: $N=tr=bk$.

Number of times any two treatments occur together in a block: $\lambda=r\,(k-1)/(t-1)$.

Sum of all N experimental values (i.e. "grand total"): G.

Sum of all r experimental values for treatment 1: T_1.

Sum of all k experimental values for block 1: B_1.

Error variance of a single experimental value when arranged in blocks of k units: σ_k^2.

It is first necessary to calculate the t quantities

$$Q_1=kT_1-B_1-B_2-\ldots-B_r,$$

etc., where the r B's for Q_1 are the totals of the r blocks containing treatment 1, etc.

The treatment effects, in terms of the change produced in a single experimental unit, are represented by the differences of the quantities

$$v_1=\frac{(t-1)}{N\,(k-1)}\,Q_1,$$

etc., the standard error of each difference being

$$\sqrt{2}\times\sqrt{\frac{k\,(t-1)}{N\,(k-1)}\,\sigma_k^2}.$$

It is best to add the general mean G/N to each of these quantities in order that their mean shall equal the general mean.

The sums of squares for blocks and treatments in the analysis of variance are as follows:

$$\text{Blocks:} \quad \frac{1}{k}(B_1^2+B_2^2+\ldots+B_b^2)-\frac{1}{N}G^2.$$

$$\text{Treatments:} \quad \frac{t-1}{Nk\,(k-1)}(Q_1^2+Q_2^2+\ldots+Q_t^2).$$

If the number of experimental units in a block is greater than one half the number of treatments the quantities Q are best replaced by

$$Q_1' = kT_1 + B_{r+1} + \ldots + B_b,$$

etc., the B's for Q_1' being those in which treatment 1 does *not* occur.

The quantities representing the treatment effects must now be diminished by

$$G\,(t-k)/N\,(k-1)$$

if their mean is to equal the general mean, and the sum of the squares of Q must be replaced by the sum of the squares *of the deviations* of Q'.

It is worth noting that the numerical factors in the above expressions can be written

$$\frac{1}{kr}\cdot\frac{1-1/t}{1-1/k}, \quad \frac{1}{r}\cdot\frac{1-1/t}{1-1/k}, \quad \frac{1}{k^2r}\cdot\frac{1-1/t}{1-1/k}.$$

The first part of each of these factors is the fraction that would be obtained if the quantities kT_1, etc., were being analysed instead of Q.

Normally estimates of the differences between blocks (allowing for possible treatment effects) are of no interest. If, however, they are required, as in the example given later, they can be calculated by deducting from each block mean the mean of the v's of the treatments in that block, or alternatively by adding the mean of the v's of the treatments not in the block.

In the special case of the arrangements which are such that the same relations hold for blocks as for treatments, i.e. in which each pair of blocks has the same number of treatments in common, the estimates of the differences between blocks may be calculated in the same way as the treatment effects, and the standard error of each difference will be the same, and equal to the standard error of differences of the treatment effects. Only a few arrangements can possibly be of this nature.

In the more general case the standard error of the difference of two block means having μ treatments in common is

$$\sqrt{2}\times\sqrt{\frac{1}{k}\left\{1+\frac{(k-\mu)\,(t-1)}{rt\,(k-1)}\right\}\sigma_k^2}.$$

The average value of μ is $k\,(r-1)/(b-1)$, and therefore the average value of the variance of the difference of the block means is

$$\bar{V}_b = \frac{2\sigma_k^2}{k}\left\{1+\frac{(t-1)\,(t-k)}{t\,(k-1)\,(b-1)}\right\}.$$

Efficiency

The expression for the standard error of a treatment comparison enables us to make a comparison of the efficiencies of symmetrical incomplete blocks and the ordinary randomized block or Latin square arrangement, provided we know the ratio of the error variances of the experimental units when arranged in blocks of size k and in blocks of size t.

The variance of the difference of two treatment means in an ordinary randomized block experiment with r replicates will be $2\sigma_t^2/r$, and therefore the ratio of this variance to that obtained with an arrangement in incomplete blocks of size k will be

$$\frac{2\sigma_t^2}{r}\Big/\frac{2k\,(t-1)}{N\,(k-1)}\,\sigma_k^2=\frac{1-1/k}{1-1/t}\cdot\frac{\sigma_t^2}{\sigma_k^2},$$

since with the same number of units the number of replicates will be the same in both arrangements. This expression is therefore a measure of the relative efficiency of the two methods. The fraction

$$\frac{1-1/k}{1-1/t}$$

may be called the *efficiency factor* of the incomplete block arrangement, and measures the loss of information when there is no reduction in error variance per unit by reduction in block size. It may be looked on as a measure of the inevitable loss inherent in the arrangement.

In agricultural field trials, and other experimental material in which there is no definite limit to the number of experimental units in a block, the ratio σ_t^2/σ_k^2 can only be properly determined by the examination of uniformity trial data, the error variances being calculated for arrangements in blocks both of t and of k units. In material in which the natural block size is definitely fixed, however, with no association between different blocks, a knowledge of the variance between and within blocks is all that is required. This can be determined from the results of actual experiments arranged in blocks of the natural size of k units.

In this latter case when treatments produce no effect any experimental value y can be regarded as made up of the sum of three parts, one constant over the whole experiment, a second α varying from block to block of k units but constant for all values in a given block, and a third β varying independently from value to value. If the variance of α is $V(\alpha)$ and of β is $V(\beta)$ it can be shown that the average variance within blocks of t units made up of numbers $l, l', l'', \ldots$ of units from different blocks of k units is equal to

$$\frac{2\,(ll'+ll''+l'l''+\ldots)}{t\,(t-1)}\,V(\alpha)+V(\beta).$$

This expression, or its mean for the various sets of $l, l', l'', \ldots$ required, will be equal to σ_t^2, and $V(\beta)$ is σ_k^2, so that σ_t^2/σ_k^2 is calculable if $V(\alpha)/V(\beta)$ (equal to ρ say) is known. The ratio of the mean square for blocks to the mean square for error in the ordinary analysis of variance of an experiment in blocks of size k will be equal to $k\rho+1$, except for errors of estimation. In order to remove bias when estimating the mean ρ from a group of experiments each such ratio should be multiplied by $(n-2)/n$, n being the number of degrees of freedom for error.

Table VI shows the values in terms of ρ of the ratio of the efficiencies for natural blocks of three, four and six, and treatments from four to ten; the first column of figures for each value of k will be recognized as the previously defined efficiency factor. The values ρ_0 of ρ for which the ratio is equal to unity are also given. These latter range from about 0·4 for blocks of three down to about 0·12 for blocks of six. With blocks of six, therefore, even slight block heterogeneity will make the method of incomplete randomized blocks more efficient; with blocks of three, on the other hand, there must be considerable block heterogeneity.

Table VI. *Ratio of the efficiencies of arrangements in incomplete and complete randomized blocks*

	$k=3$		$k=4$		$k=6$	
	Ratio	ρ_0	Ratio	ρ_0	Ratio	ρ_0
4	$0{\cdot}889+0{\cdot}444\rho$	0·25				
5	$0{\cdot}833+0{\cdot}528\rho$	0·32	$0{\cdot}938+0{\cdot}375\rho$	0·17		
6	$0{\cdot}8\ \ +0{\cdot}48\rho$	0·42	$0{\cdot}9\ \ +0{\cdot}48\rho$	0·21		
7	$0{\cdot}778+0{\cdot}556\rho$	0·40	$0{\cdot}875+0{\cdot}531\rho$	0·24	$0{\cdot}972+0{\cdot}278\rho$	0·10
8	$0{\cdot}762+0{\cdot}580\rho$	0·41	$0{\cdot}857+0{\cdot}490\rho$	0·29	$0{\cdot}952+0{\cdot}408\rho$	0·12
9	$0{\cdot}75\ \ +0{\cdot}562\rho$	0·44	$0{\cdot}844+0{\cdot}562\rho$	0·28	$0{\cdot}938+0{\cdot}469\rho$	0·13
10	$0{\cdot}741+0{\cdot}593\rho$	0·44	$0{\cdot}833+0{\cdot}593\rho$	0·28	$0{\cdot}926+0{\cdot}521\rho$	0·14

Possible arrangements

The condition that every two treatments occur together in a block an equal number of times is satisfied if the blocks are chosen so as to include every possible grouping of the t treatments k at a time. In this case

$$b={}_tC_k,\quad r={}_{t-1}C_{k-1},\quad \lambda={}_{t-2}C_{k-2},$$

where ${}_tC_k$ is the number of combinations of t things k at a time, and equals

$$t\,(t-1)\,\ldots\,(t-k+1)/k\,(k-1)\,\ldots\,2.1.$$

If f is the highest common factor of the above values of b, r and λ lower limits of b, r and λ are given by $1/f$ of these values, since all must be whole numbers. It does not follow, however, that any arrangement exists for these lower limits.

Table VII. *Lower limits to the number of replications* (r) *in symmetrical incomplete randomized block experiments of various sizes*

Number of treatments (t)	Number of units per block (k)										
	2	3	4	5	6	7	8	9	10	11	12
25	24*	(12)	(8)	6	(24)	(28)	(24)	(9)	(8)	(132)	(24)
24	23*	(23)	(23)	(115)	(23)	(161)	(23)	(69)	(115)	(253)	(23)
23	22*	(33)	(44)	(55)	(66)	(77)	(88)	(99)	(110)	(11)	
22	21*	(21)	(14)	(105)	(21)	(7)	(12)	(63)	(35)	(21)	
21	20*	(10)	(20)	5	[4]	(10)	(40)	(15)	(20)		
20	19*	(57)	(19)	(19)	(57)	(133)	(38)	(171)	(19)		
19	18*	(9)	(12)	(45)	(18)	(21)	(72)	(9)			
18	17*	(17)	(34)	(85)	(17)	(119)	(68)	(17)			
17	16*	(24)	(16)	(20)	(48)	(56)	(16)				
16	15*	(15)	5	(15)	[3] 6	(35)	15				
15	14*	(7)	(28)	(7)	(14)	(7)					
14	13*	(39)	(26)	(65)	(39)	(13)					
13	12*	6	4	(15)	(12)						
12	11*	(11)	(11)	(55)	(11)						
11	10*	(15)	(20)	5							
10	9*	9	6	9							
9	8*	4	8								
8	7*	21*	7								
7	6*	3									
6	5*	5									
5	4*										
4	3*										

Curved brackets ( ) indicate that the possibility of an arrangement has not been investigated, square brackets [ ] that no arrangement exists.

An arrangement is shown in Table VIII for each of the unbracketed values, except those, marked by an asterisk (*), which involve all possible combinations, and $t=16$, $k=8$, $r=15$. The arrangements for 31, 49, etc., treatments, described in the text on p. 134, but not shown in Table VII or VIII, should also be noted.

When k is greater than $\frac{1}{2}t$ lower limits and arrangements can be derived as indicated in the text.

Table VII gives these lower limits of r for all t up to twenty-five and all k not greater than $\frac{1}{2}t$. The investigation of whether any arrangement is possible for any given values of t, k and r involves intricate and extensive combinatorial problems which have not in general been solved, but certain of the simpler cases have been investigated. For numbers of treatments not exceeding ten an arrangement has been found for every value of k. For numbers of treatments greater than ten only those cases in which the lower limit of r is not greater than six have been examined. Two of these, $t=16$, $k=6$, $r=3$, and $t=21$, $k=6$, $r=4$ were found to be impossible, but the former yielded an arrangement, discussed below, for $r=6$. Table VIII gives the various arrangements which have been found (other than those in which all possible groupings are necessary).

Table VIII. *Experimental arrangements*

(*a*) Arrangements symmetrical for blocks as well as for treatments

Seven treatments: 7 blocks of 3 (3-fold)

a	*b*	*c*	*d*	*e*	*f*	*g*
+	+	+	.	.	.	.
+	.	.	+	+	.	.
+	.	.	.	.	+	+
.	+	.	+	.	+	.
.	+	.	.	+	.	+
.	.	+	+	.	.	+
.	.	+	.	+	+	.

Eleven treatments: 11 blocks of 5 (5-fold)

a	*b*	*c*	*d*	*e*	*f*	*g*	*h*	*i*	*j*	*k*
+	+	+	+	+	.	.	.	.	.	.
+	+	.	.	.	+	+	+	.	.	.
+	.	+	.	.	+	.	.	+	+	.
+	.	.	+	.	.	+	.	+	.	+
+	.	.	.	+	.	.	+	.	+	+
.	+	+	.	.	.	+	.	.	+	+
.	+	.	+	.	.	.	+	+	+	.
.	+	.	.	+	+	.	.	+	.	+
.	.	+	+	.	+	.	+	.	.	+
.	.	+	.	+	.	+	+	+	.	.
.	.	.	+	+	+	+	.	.	+	.

Thirteen treatments: 13 blocks of 4 (4-fold)

a	*b*	*c*	*d*	*e*	*f*	*g*	*h*	*i*	*j*	*k*	*l*	*m*
+	+	+	+	.	.	.	.	.	.	.	.	.
+	.	.	.	+	+	+	.	.	.	.	.	.
+	.	.	.	.	.	.	+	+	+	.	.	.
+	.	.	.	.	.	.	.	.	.	+	+	+
.	+	.	.	+	.	.	+	.	.	+	.	.
.	+	.	.	.	+	.	.	+	.	.	+	.
.	+	.	.	.	.	+	.	.	+	.	.	+
.	.	+	.	+	.	.	.	+	.	.	.	+
.	.	+	.	.	+	.	.	.	+	+	.	.
.	.	+	.	.	.	+	+	.	.	.	+	.
.	.	.	+	+	.	.	.	.	+	.	+	.
.	.	.	+	.	+	.	+	.	.	.	.	+
.	.	.	+	.	.	+	.	+	.	+	.	.

Sixteen treatments: 16 blocks of 6 (6-fold)

a	*b*	*c*	*d*	*e*	*f*	*g*	*h*	*i*	*j*	*k*	*l*	*m*	*n*	*o*	*p*
+	+	+	+	+	+	.	.	.	.	.	.	.	.	.	.
+	+	.	.	.	.	+	+	+	+	.	.	.	.	.	.
+	.	+	.	.	.	+	.	.	.	+	+	+	.	.	.
+	.	.	+	.	.	.	+	.	.	+	.	.	+	+	.
+	.	.	.	+	.	.	.	+	.	.	+	.	+	.	+
+	.	.	.	.	+	.	.	.	+	.	.	+	.	+	+
.	+	+	.	.	.	+	.	.	.	.	.	.	+	+	+
.	+	.	+	.	.	.	+	.	.	.	+	+	.	.	+
.	+	.	.	+	.	.	.	+	.	+	.	+	.	+	.
.	+	.	.	.	+	.	.	.	+	+	+	.	+	.	.
.	.	+	+	.	.	.	.	+	+	+	.	.	.	.	+
.	.	+	.	+	.	.	+	.	+	.	+	.	.	+	.
.	.	+	.	.	+	.	+	+	.	.	.	+	+	.	.
.	.	.	+	+	.	+	.	.	+	.	.	+	+	.	.
.	.	.	+	.	+	+	.	+	.	.	+	.	.	+	.
.	.	.	.	+	+	+	+	.	.	+	.	.	.	.	+

Table VIII (*cont.*)

Twenty-one treatments: 21 blocks of 5 (5-fold)

a	*b*	*c*	*d*	*e*	*f*	*g*	*h*	*i*	*j*	*k*	*l*	*m*	*n*	*o*	*p*	*q*	*r*	*s*	*t*	*u*
+	+	+	+	+	.	.	.	.	.	.	.	.	.	.	.	.	.	.	.	.
+	.	.	.	.	+	+	+	+	.	.	.	.	.	.	.	.	.	.	.	.
+	.	.	.	.	.	.	.	.	+	+	+	+	.	.	.	.	.	.	.	.
+	.	.	.	.	.	.	.	.	.	.	.	.	+	+	+	+	.	.	.	.
+	.	.	.	.	.	.	.	.	.	.	.	.	.	.	.	.	+	+	+	+
.	+	.	.	.	+	.	.	.	+	.	.	.	+	.	.	.	+	.	.	.
.	+	.	.	.	.	+	.	.	.	+	.	.	.	+	.	.	.	+	.	.
.	+	.	.	.	.	.	+	.	.	.	+	.	.	.	+	.	.	.	+	.
.	+	.	.	.	.	.	.	+	.	.	.	+	.	.	.	+	.	.	.	+
.	.	+	.	.	+	.	.	.	.	.	.	+	.	+	.	.	.	.	+	.
.	.	+	.	.	.	+	.	.	.	.	+	.	+	.	.	.	.	.	.	+
.	.	+	.	.	.	.	+	.	.	+	.	.	.	.	.	+	+	.	.	.
.	.	+	.	.	.	.	.	+	+	.	.	.	.	.	+	.	.	+	.	.
.	.	.	+	.	+	.	.	.	.	+	.	.	.	.	+	.	.	.	.	+
.	.	.	+	.	.	+	.	.	+	.	.	.	.	.	.	+	.	.	+	.
.	.	.	+	.	.	.	+	.	.	.	.	+	+	.	.	.	.	+	.	.
.	.	.	+	.	.	.	.	+	.	.	+	.	.	+	.	.	+	.	.	.
.	.	.	.	+	+	.	.	.	.	.	+	.	.	.	.	+	.	+	.	.
.	.	.	.	+	.	+	.	.	.	.	.	+	.	.	+	.	+	.	.	.
.	.	.	.	+	.	.	+	.	+	.	.	.	.	+	.	.	.	.	.	+
.	.	.	.	+	.	.	.	+	.	+	.	.	+	.	.	.	.	.	+	.

(*b*) Other arrangements

Six treatments: 10 blocks of 3 (5-fold). See Introduction

Eight treatments: 14 blocks of 4 (7-fold)

a b c d	*a b e f*	*a c e g*	*a b g h*	*a c f h*	*a d e h*	*a d f g*
e f g h	*c d g h*	*b d f h*	*c d e f*	*b d e g*	*b c f g*	*b c e h*

Nine treatments: 12 blocks of 3 (4-fold)

a b c	*a d g*	*a e i*	*a f h*
d e f	*b e h*	*b f g*	*b d i*
g h i	*c f i*	*c d h*	*c e g*

Nine treatments: 18 blocks of 4 (8-fold)

a b c d	*a c e g*	*a d h i*	*b d e i*	*b e f h*	*c d e f*
a b e f	*a d f h*	*a e g i*	*b f g i*	*c e h i*	*c f g h*
a b g h	*a c f i*	*b c h i*	*b c d g*	*d f g i*	*d e g h*

Ten treatments: 30 blocks of 3 (9-fold)

a b c	*a f h*	*b d j*	*b h j*	*c h i*	*d g h*
a b d	*a g i*	*b e h*	*c d g*	*c i j*	*e f j*
a c e	*a h j*	*b e i*	*c d h*	*d e i*	*e g h*
a d f	*a i j*	*b f g*	*c e f*	*d e j*	*f g j*
a e g	*b c f*	*b g i*	*c g j*	*d f i*	*f h i*

Ten treatments: 15 blocks of 4 (6-fold)

a b c d	*a d i j*	*b c f i*	*b g h i*	*c d e h*
a b e f	*a e g i*	*b d g j*	*c e i j*	*d e f g*
a c g h	*a f h j*	*b e h j*	*c f g j*	*d f h i*

Ten treatments: 18 blocks of 5 (9-fold)

a b c d e	*a c f h i*	*a e g i j*	*b e f h j*	*c e f h i*
a b c f g	*a c g h j*	*b c d h j*	*b f g i j*	*d e f g h*
a b d f i	*a d e f j*	*b c e i j*	*c d f g j*	
a b e g h	*a d h i j*	*b d g h i*	*c d e g i*	

Thirteen treatments: 26 blocks of 3 (6-fold)

a b c	*a j k*	*b f k*	*c e j*	*d f j*	*e g l*	*h j m*
a d e	*a l m*	*b g j*	*c f l*	*d g m*	*e i k*	*i j l*
a f g	*b d i*	*b h l*	*c g i*	*d k l*	*f i m*	
a h i	*b e m*	*c d h*	*c k m*	*e f h*	*g h k*	

Table VIII (*cont.*)

Sixteen treatments: 20 blocks of 4 (5-fold)

a b c d	a e i m	a f k p	u h j o	a g l n
e f g h	b f j n	b g l m	b e k p	b h i o
i j k l	c g k o	c h i n	c f l m	c e j p
m n o p	d h l p	d e j o	d g i n	d f k m

Twenty-five treatments: 30 blocks of 5 (6-fold)

a b c d e	a f k p u	a g m s y	a h o q x	a i l t w	a j n r v
f g h i j	b g l q v	b h n t u	b i k r y	b j m p x	b f o s w
k l m n o	c h m r w	c i o p v	c j l s u	c f n q y	c g k t x
p q r s t	d i n s x	d j k q w	d f m t v	d g o r u	d h l p y
u v w x y	e j o t y	e f l r x	e g n p w	e h k s v	e i m q u

In the case of blocks of two, i.e. pairs of experimental units, all possible pairs of treatments must be taken; there are $\frac{1}{2}t\,(t-1)$ such pairs, giving a minimum of $t-1$ replications. Arrangements for blocks containing one less unit than the number of treatments are formed by omitting every treatment an equal number of times, and arrangements for blocks containing two less units are formed by omitting every treatment pair an equal number of times.

In general the minimum number of replications of t treatments in blocks of $t-k$ units is equal to the minimum number of replications in blocks of k units multiplied by $(t-k)/k$, an arrangement for blocks of k units being convertible into one for blocks of $t-k$ units by replacing every block by its complement, i.e. by a block containing all the treatments missing from the original block.

The five arrangements in which t equals b, shown in the first part of Table VIII, have the additional property already referred to that every pair of blocks has the same number of treatments in common. No arrangement having this property is possible for the values of k, t and r for which arrangements are shown in the second part of the table, since in no one of these cases has $\bar{\mu}$ an integral value.

Three series of values from Table VII are worth comment. Arrangements for the series in which $\lambda=1$ and $k=r-1$, and the series in which $\lambda=1$ and $k=r$, can both be derived from completely orthogonalized Latin squares (i.e. hyper-Graeco-Latin squares) of side $r-1$. The values of t corresponding to the first few values of r for these series are:

	r	3	4	5	6	7	8	9	10
t	$\lambda=1,\ k=r-1$:	4	9	16	25	[36]	49	64	81
	$\lambda=1,\ k=r$:	7	13	21	31	[43]	57	73	91

It is known that completely orthogonalized squares exist when the side is a prime number and also for sides 4, 8 and 9(2). It is also known that no such square of side 6 exists(3). Higher non-primes have not been investigated.

The structure of the first series is easily seen. If the treatments are set out in the form of a square and on this a hyper-Graeco-Latin square is superimposed the first set of blocks is formed from the rows of this square, the second from the columns, the third from treatments having the same Latin letter, the fourth from those having the same Greek letter, and so on. (In the case of primes the selection for the third, fourth, fifth, etc., sets is made by moving one, two, three, etc., columns in proceeding from one row to the next.) The arrangements for nine, sixteen and twenty-five treatments are shown in Table VIII.

The structure of the second series is exemplified in Table VIII by the arrangements for seven, thirteen, and twenty-one treatments. The arrangement for thirteen treatments will serve as an example. The first four treatments a, b, c, d form the first block. The remaining nine may then be arranged in the form of a square:

e	f	g
h	i	j
k	l	m

The next set of three blocks $a\,e\,f\,g$, etc., is formed by taking a with each of the rows of this square. The third set is formed by taking b with each of the columns, the fourth and fifth by taking c and d respectively with the two sets of three groups derived from the right and left diagonal groupings.

The series $\lambda = 2$ and $k = r$ is also interesting, and I am indebted to Prof. Fisher for the elucidation of its structure. The first few values of this series are:

r:	3	4	5	6	7	8	9	10
t:	4	7	11	16	(22)	(29)	(37)	(46)

For the first two values k is greater than $\frac{1}{2}t$, the arrangement for $t = 7$ being the complement of that already considered for $k = 3$ and $t = 7$. Arrangements for eleven and sixteen treatments are given in Table VIII.

We will first consider the arrangement for eleven treatments. If the first five rows and the first five columns are written as shown, any one of the treatments f to k is defined by its conjunction with two of the treatments b to e in blocks 2–5. Thus f has b and c as key letters. In blocks 6–11, therefore, f must occur once with b, once with c, and twice with d and e. It must also occur once with g, h, i, and j, all of which has b or c as one of their key letters, and twice with k, which has neither b nor c as a key letter. Moreover, each of the blocks 6–11 can itself be defined by two of the key letters b to e.

Consider now the following diagram for $f\,(b, c)$:

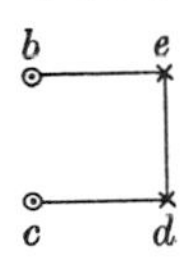

which is such that one line passes through b and c, and two lines through d and e. The lines be, cd, and de are to be taken to represent the three of the blocks 6–11 in which treatment f occurs. An arrangement will be possible if a set of diagrams of this type can be found such that every pair of diagrams having a key letter in common have one line in common and every other pair have two lines in common, for the treatments f to k will then each have two conjunctions. It is easy to see that

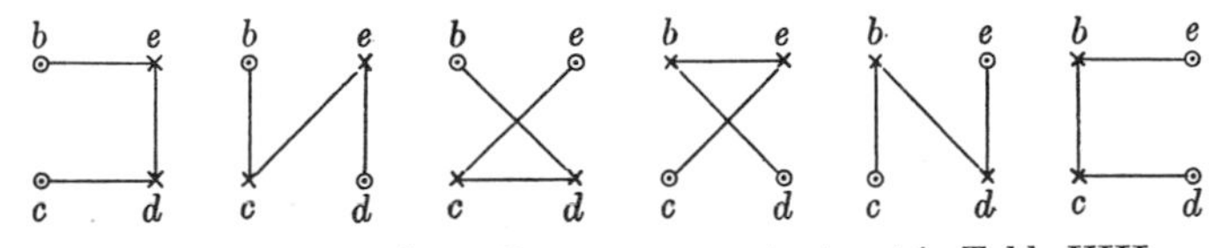

is such a set. This set corresponds to the arrangement set out in Table VIII.

The case of sixteen treatments involves five points, and a system of diagrams of the type

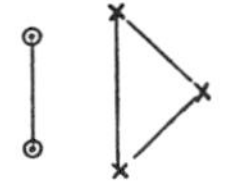

fulfils the required conditions. The higher members of this series have not been exhaustively investigated; the possible alternatives here become numerous. It should also be noted that there are other ways of arranging the first r rows which might conceivably yield arrangements, though such possibilities can be immediately excluded for values of r up to 7.

Example

In a preliminary report (1) on an experiment in which different generations of rats were being tested for ability to escape from a tank by a given route Prof. A. E. Crew publishes the scores of rats of various litters from five lines, A, B, C, G, K. The numbers of young in each litter are shown in Table IX.

Crew was concerned with the inheritance of ability in this one test, and tested every rat of every litter. Here we wish to consider the design of a suitable experiment for measuring the effects of variations in the test, or of different treatments of the rats prior to the test, e.g. the use of different drugs. We will assume, moreover, that the experimenter is interested also in the mean value for each litter of whatever ability he is measuring and that the governing factor in limiting the size of the experiment is the number of rats tested. This situation is very likely to arise when ample funds are available but sufficiently skilled assistance is hard to find; in any case the cost of rats destroyed at birth can hardly be of much consequence.

From Table IX we see that there are only four litters out of forty-three with more than seven rats and only nine with more than six rats. If, therefore, more than six variants are to be tested an ordinary randomized block arrangement with each block formed of a complete litter will not be of much use. Even if there are only six variants many litters will be rejected, and therefore no information on inheritance will be obtained from these litters.

Table IX

	A	B	C	G	K
F_1	6, (6), (5)	4, 1	5	3, 3, 4, 7	6, (7)
F_2	6, 6, 7, 5	3, 1, 2	5, 4, 4, 7, 9, 1	4, 6	7
F_3	5, 5, 10, 6, 9	3, 4, 5	5, 10, 5, 7*	5, 4	—
F_4	5	—	5*	4, 6	—

Brackets indicate litter died or was eaten before testing, an asterisk (*) that testing was not complete when the report was written.

The effect of this rejection is even more serious than it appears at first sight, for the different lines exhibit very different levels of fertility, and line B would be entirely excluded from the tests if litters of six were demanded. In this test there is an association between fertility and success in the test which would remain unobserved if only rats of high fertility were used. Such differential fertility is usually of considerable interest: in this case it may account for part at least of the remarkable changes observed by Prof. McDougall.

We are therefore forced to consider the use of randomized blocks made up of rats from more than one litter, or alternatively incomplete randomized blocks. If four rats from each litter are tested all litters except five of line B (which will require special treatment) will be included. We shall not, of course, obtain as much information on the large litters as we should if every rat were tested, but we shall be able to spread the tests over a larger number of litters (thus including, if we wish, more lines). In this connection it should be remembered that when comparing a litter of four rats with a litter of eight we only obtain one-third more information if all eight of the second litter are tested instead of only four.

In order to see the results that may be expected from material of this type, Crew's data were analysed as a uniformity trial, taking a random selection of four rats from each litter, and excluding those litters with less than four rats. Square roots of the test scores were analysed, the analysis of variance given in Table X being obtained.

Table X. *Analysis of variance, all litters, four rats from each litter*

	D.F.	Sum of squares	Mean square
Between lines	4	54·99	13·748
Between litters within lines	28	201·22	7·186
Within litters	99	309·71	3·128
Total	131	565·92	4·320

The additional variance due to litters within the same line, though quite appreciable, is not very large, the value for ρ being

$$\rho = \frac{\frac{1}{4}(7{\cdot}186 - 3{\cdot}128)}{3{\cdot}128} = 0{\cdot}324.$$

The values given in Table VI show that the use of ordinary randomized blocks, made up as far as necessary from two or more litters, will be about as efficient as incomplete blocks. With five treatments incomplete blocks are 6 per cent. more efficient, with ten treatments $2\frac{1}{2}$ per cent. By the use of incomplete blocks we shall, however, avoid the necessity of using residuals to calculate the litter differences. The reduction in the apparent differences between litters by the use of residuals is quite appreciable when the number of replications r of each treatment is small, the differences of two litters having no treatment in common being reduced (on the average) by $1/r$ times the true difference.

We will illustrate the computations by superimposing an arrangement of seven dummy variants on the first seven litters of line C, which give the four replications required for a symmetrical arrangement.

Table XI shows the square roots of these litters, and also the random arrangement obtained. The set of treatments assigned to each litter was determined (using the arrangement given in Table VIII) by random choice. The treatment totals and the quantities Q are shown in Table XII. It will be noted that the sum of all the Q's is seven times the grand total of all the values. The quantities in the last line of Table XII are the adjusted treatment means, being equal to $\frac{1}{14}Q - 4{\cdot}318$. The differences of these quantities represent the estimates of the effects of the treatments on a single rat.

Table XI. *Square roots of scores, line C*

	Litter						
	1	2	3	4	5	6	7
	(*b*) 0·0	(*g*) 2·4	(*f*) 2·4	(*b*) 3·6	(*g*) 4·7	(*b*) 5·1	(*e*) 7·5
	(*g*) 2·2	(*d*) 2·6	(*c*) 2·4	(*a*) 5·0	(*f*) 5·3	(*d*) 4·7	(*a*) 2·2
	(*f*) 3·3	(*a*) 6·2	(*d*) 3·0	(*f*) 5·1	(*e*) 5·5	(*e*) 7·6	(*g*) 2·6
	(*c*) 5·0	(*b*) 1·4	(*a*) 4·0	(*e*) 6·5	(*d*) 6·6	(*c*) 9·6	(*c*) 4·4
Total	10·5	12·6	11·8	20·2	22·1	27·0	16·7
Q'	103·4	115·0	96·3	131·0	137·3	153·4	109·9
Adjusted litter means	3·07	3·90	2·56	5·04	5·49	6·64	3·53

Table XII. *Treatment totals and adjusted means*

	Treatment							Total or mean
	a	*b*	*c*	*d*	*e*	*f*	*g*	
Total	17·4	10·1	21·4	16·9	27·1	16·1	11·9	120·9
Q	129·2	91·0	140·5	115·0	143·3	120·7	106·6	846·3
Adjusted mean	4·91	2·18	5·72	3·90	5·92	4·30	3·30	4·318

The analysis of variance is given in Table XIII. The sum of squares between litters is computed in the ordinary manner. The sum of squares for treatments is the sum of squares of the deviations of the Q's multiplied by $6/28.4.3$ or $1/56$. The error sum of squares is obtained by subtraction.

Table XIII. *Analysis of variance*

	Degrees of freedom	Sum of squares	Mean square
Between litters	6	56·12	9·353
Treatments	6	37·28	6·213 } 3·212
Error	15	30·18	2·012 } 3·212
Total	27	123·58	4·577

Fitting treatments before litters (test for litters):

	Degrees of freedom	Sum of squares	Mean square
Between treatments	6	48·86	8·143
Litters	6	44·54	7·423

The arrangement is notable in that the analysis shows a significant difference between the dummy treatments, for $z = 0{\cdot}564$, whereas the 5 per cent. point is 0·513. That this is due to a genuine chance conjunction of one treatment with rats having good performances (low scores) relative to their litter mates, and not to any defect of the statistical analysis, is indicated by examination of Table XI. Treatment (b) has much the lowest score in three of the four litters in which it occurs, and a score nearly equal to the lowest in the fourth. The chance of any one of the seven treatments having either all four lowest or all four highest scores of the litters concerned is approximately $2 \times 7/4^4$ or $1/18$.

The standard error of the difference of any two of the adjusted treatment means is equal to

$$\sqrt{2 \times \frac{4.6}{28.3} \times 2{\cdot}012} = 1{\cdot}07.$$

In the arrangement of seven treatments in blocks of four every pair of blocks has the same number of treatments in common, so that the differences between litters can here be estimated in precisely the same manner as the differences between treatments. Quantities Q', similar to Q, and adjusted litter means, have been calculated and are shown in Table XI. The standard errors are the same as those already calculated.

Had an arrangement which was not reciprocally symmetrical been used the differences between litters could only have been computed by applying corrections for the treatment differences. The adjusted mean for litter 1, for instance, is given by

$$\tfrac{1}{4}(10{\cdot}5 - 2{\cdot}18 - 5{\cdot}72 - 4{\cdot}30 - 3{\cdot}30) + 4{\cdot}318 = 3{\cdot}07,$$

or by

$$\tfrac{1}{4}(10{\cdot}5 + 4{\cdot}91 + 3{\cdot}90 + 5{\cdot}92 - 3 \times 4{\cdot}318) = 3{\cdot}07.$$

Summary

The paper describes a general method of arranging replicated experiments in randomized blocks when the number of treatments to be compared is greater than the number of experimental units in a block. This new type of arrangement, for which the name of *symmetrical incomplete randomized blocks* is proposed, is such that every two treatments occur together in a block the same number of times. This restriction enables estimates of the treatment effects and of the experimental error to be obtained expeditiously by the ordinary procedure of the analysis of variance. Estimates of block differences can also be obtained if required. The special case in which the blocks are formed of pairs of experimental units is capable of specially simple treatment. The method of symmetrical incomplete randomized blocks is likely to be of most use in cases in which the experimental material naturally divides itself into groups, such as litters of experimental animals, containing numbers less than the number of treatments that it is desired to test, especially if the differences between these natural groups are of interest.

The necessary formulae are presented and their application illustrated by numerical examples, one based on the numbers of local lesions produced by a virus on half leaves of susceptible plants, the other on the scores of rats in a discrimination test. The minimum number of replications required for different numbers of treatments and block sizes is discussed, and actual arrangements are given for the cases likely to be of general utility. A short discussion of the relative efficiency of an arrangement of this type and an arrangement in ordinary randomized blocks is also included.

REFERENCES

(1) Crew, F. A. E. (1932). "Inheritance of educability." *Proc. Sixth Int. Congr. Genet.* **1**, 121–34.
(2) Fisher, R. A. (1936). *The Design of Experiments.* 2nd Edn., Chap. v.
(3) Fisher, R. A. & Yates, F. (1934). "The 6 × 6 Latin squares." *Proc. Camb. phil. Soc.* **30**, 492–507.
(4) Yates, F. (1936). "Incomplete Latin squares." *J. agric. Sci.* **26**, 301–15.
(5) —— (1936). "A new method of arranging variety trials involving a large number of varieties." *J. agric. Sci.* **26**, 424–55.

Paper VIII

THE RECOVERY OF INTER-BLOCK INFORMATION IN BALANCED INCOMPLETE BLOCK DESIGNS

From ANNALS OF EUGENICS
Volume 10, Part 4, pp. 317–325, 1940

Author's Note

Paper VIII investigates the recovery of interblock information in balanced incomplete block designs. Classical least squares deals only with the cases in which all the observations are subject to the same error, or in which the error variances are in known ratios (weighted observations). Fisher, in the analysis of split plot designs, showed how a hierarchy of errors could be postulated and estimated, the treatment effects being divided into groups, each with its appropriate error. This paper takes the idea further, in that separate estimates of some or all of the treatment effects are obtained from inter- and intra-block comparisons, to which separate estimatable errors can be assigned, and these estimates are then combined. As such it represents an interesting non-standard application of least squares.

1969 FRANK YATES

THE RECOVERY OF INTER-BLOCK INFORMATION IN BALANCED INCOMPLETE BLOCK DESIGNS

By F. YATES

1. Introduction

Incomplete block and quasi-factorial designs of various kinds were first introduced by the author a few years ago (1936*a*, *b*). All these designs have the property that the number of varieties (or treatments) included in each block is smaller than the total number to be tested. There is consequently a gain in precision due to the use of smaller blocks, at the expense of loss of information on those varietal comparisons which are confounded with blocks. In the original papers only the complete elimination of inter-block differences was considered. These inter-block comparisons will, however, contain an appreciable amount of information, amounting in the limiting case, when the inter-block and intra-block comparisons are of equal accuracy, to a fraction $1-E$ of the total information, where E is the efficiency factor.

The recovery of this information has already been discussed for three-dimensional quasi-factorial designs (1939) and also for lattice (quasi-Latin) squares (1940). The present paper contains a similar discussion of balanced incomplete block designs. The case of two-dimensional quasi-factorial designs is to be dealt with in a publication by the Statistical Department of Iowa State College.

2. Estimates of the varietal differences

If v varieties are arranged in b blocks containing k varieties each, there being r replicates of each variety, the condition of balance will be fulfilled if each pair of varieties occurs together an equal number λ of times. A catalogue of possible designs and known solutions is given by Fisher & Yates (1938).

The following relations hold:

$$vr = kb,$$

$$(v-1)\lambda = (k-1)r.$$

The efficiency factor E, defined as the fraction of the total information contained in the intra-block comparisons, when inter-block and intra-block comparisons are of equal accuracy, is given by

$$E = \frac{1-1/k}{1-1/v} = \frac{v\lambda}{rk}.$$

Let V_s be the sum of all the yields of variety s, T_s the sum of all the block totals of blocks containing variety s, T'_s the sum of all the remaining block totals, and G the total yield of all plots. Then, as has been shown previously (1936*a*), the estimates of the varietal differences derived from the intra-block comparisons are obtained from the quantities

$$Q_s = V_s - T_s/k,$$

or, as is more convenient when $k > \frac{1}{2}v$,

$$Q'_s = V_s + T'_s/k.$$

The actual differences in units of the total yield of the r replicates are given by the differences of

$$Q_s/E \quad \text{or} \quad Q'_s/E,$$

the sum of the first set being zero, and the second set $rG/\lambda v$.

The error variance of these latter sets of quantities is r/Ew, where $1/w$ is the intra-block error variance.

The estimates of the varietal differences derived from the inter-block comparisons are similarly given by the differences of

$$rT_s/(r-\lambda) \quad \text{or} \quad rT'_s/(r-\lambda),$$

in units of the total yield of r replicates. The error variance of these sets of quantities is $kr^2/(r-\lambda)\,w'$, where $1/w'$ is the error variance of the inter-block comparisons, in units of a single plot.

If the weights w and w' are known, the most efficient estimates of the varietal differences will be given by the differences of the weighted means:

$$Y_s = \frac{\dfrac{Q_s}{E}\dfrac{Ew}{r} + \dfrac{rT_s}{r-\lambda}\dfrac{(r-\lambda)\,w'}{kr^2}}{\dfrac{Ew}{r} + \dfrac{(r-\lambda)\,w'}{kr^2}}.$$

The quantities Y_s may be termed the partially adjusted yields (totals of r replicates). If G is the total yield of all plots, Y_s may be written (after the addition of a quantity $\mu\,(k-1)\,G$) in the form

$$Y_s = V_s + \mu W_s,$$

where

$$W_s = (v-k)\,V_s - (v-1)\,T_s + (k-1)\,G,$$

and

$$\mu = \frac{w-w'}{wv(k-1)+w'(v-k)}.$$

The error variance of the Y's is

$$\frac{1}{\dfrac{Ew}{r} + \dfrac{(r-\lambda)\,w'}{kr^2}} = \frac{kr(v-1)}{wv(k-1)+w'(v-k)}.$$

3. The analysis of variance

The structure of the analysis of variance is a little complicated. The residual sum of squares for intra-block error may be calculated (*a*) by deducting the sum of squares for blocks (ignoring varieties) and for varieties (eliminating blocks) from the total sum of squares, or alternatively (*b*) by deducting the sum of squares for varieties (ignoring blocks) and for blocks (eliminating varieties).

As has previously been shown, the sum of squares for varieties (eliminating blocks) is derived from the sum of the squares of the quantities Q, with divisor rE.

The sum of squares for blocks (eliminating varieties) splits into two parts. The first, corresponding to $v-1$ degrees of freedom, is affected by varietal differences and is derived from the sum of the squares of the deviations of the quantities W_s, with divisor $rv\,(v-k)\,(k-1)$. The second, corresponding to $b-v$ degrees of freedom, is unaffected by varietal differences, and represents pure inter-block error. This latter sum of squares can best be computed by taking the difference of the total sum of squares for blocks (ignoring varieties) and the component of this sum of squares which is affected by varietal differences, this being given by the sum of squares of the deviations of T_s, with divisor $k(r-\lambda)$.

Table 1 shows these relations in tabular form. In this table dev^2 indicates the sum of the squares of the deviations, y the individual yields and B the block totals. By calculating both forms of the analysis a complete check is obtained, except for the total sum of squares and for the total sum of squares for blocks (ignoring varieties).

Table 1. *Structure of analysis of variance*

Method (*a*)	D.F.	S.S. (*a*)		S.S. (*b*)	Method (*b*)
Blocks (ignoring varieties):					Blocks (eliminating varieties):
Varietal component	$v-1$	$\dfrac{dev^2\ T}{k(r-\lambda)}$		$\dfrac{dev^2\ W}{rv(v-k)(k-1)}$	Varietal component
Remainder	$b-v$	†	→	†	Remainder
Total	$b-1$	$\dfrac{dev^2\ B^*}{k}$		†	Total
Varieties (eliminating blocks)	$v-1$	$\dfrac{dev^2\ kQ}{k^2\ rE}$		$\dfrac{dev^2\ V}{r}$	Varieties (ignoring blocks)
Intra-block error	$rv-v-b+1$	†	⟷	†	Intra-block error
Total	$rv-1$	$dev^2\ y^*$	⟷	$dev^2\ y^*$	Total

* Requires checking. † Calculated by addition or subtraction.

4. Estimation of the relative weights

If the intra-block error variance is B, and the error variance of block totals is $k(kA+B)$, the expectations of the mean squares corresponding to the components of the sum of squares for varieties (eliminating blocks) are shown in Table 2.

Table 2. *Expectations of mean squares for blocks (eliminating varieties)*

	D.F.	Expectation
Varietal component	$v-1$	$EkA+B$
Remainder	$b-v$	$kA+B$
Total	$b-1$	$\dfrac{bk-v}{b-1}A+B$

The factor E in the first of the above expressions is derived as follows. If for any pair of varieties s and s' the coefficients of each plot yield in the difference $W_s-W_{s'}$ are written down, and summed by blocks, it will be found that $(r-\lambda)$ of these sums equal $v(k-1)$ and $(r-\lambda)$ equal $-v(k-1)$, the remainder being zero. Utilizing the divisor given in Table 1 (which is itself one-half the sum of the squares of these coefficients), we obtain as the coefficient of A

$$\frac{(r-\lambda)\,v^2(k-1)^2}{rv(v-k)\,(k-1)} = Ek.$$

As has been pointed out previously (Yates, 1939), it is sufficient, for the purpose of estimating w and w', to equate the expectation in terms of w and w' of the mean square for all the $b-1$ degrees of freedom for blocks (eliminating varieties) with the actual mean square M'', say. If the mean square for intra-block error is M, we obtain the equation

$$w = \frac{1}{M}, \quad w' = \frac{v(r-1)}{k(b-1)\,M''-(v-k)\,M}.$$

Since w may ordinarily be assumed to be greater than w' it will be sufficient, if M'' is less than M, to take w' as equal to w, i.e. to use the unadjusted yields as the final estimates.

Since M'' is frequently based on a somewhat small number of degrees of freedom, there is of course some inaccuracy in the estimated weights. The effect of this inaccuracy on the accuracy of the weighted estimates has been investigated in various extreme cases (Yates, 1939, 1940; and Cochran, unpublished material). The results obtained are summarized in Table 3.

Table 3. *Loss of information due to inaccuracies of weighting*

(*a*) Particulars of cases investigated

Case	Type of design	Replications	Degrees of freedom		Expectation of block M.S.		Efficiency factor	Reference to literature
			Blocks	Error	Actual	Unconfounded		
1	5 × 5 lattice	2	8	16	$\frac{5}{2}A+B$	$5A+B$	0·75	Cochran (unpublished)
2	4 × 4 triple lattice	3	9	21	$\frac{8}{3}A+B$	$4A+B$	0·769	Cochran (unpublished)
3	3 × 3 × 3 lattice	3	24	28	$2A+B$	$3A+B$	0·591	Yates (1939)
4	5 × 5 lattice squares	3	12	24	$\frac{10}{3}A+B$	$5A+B$	0·667	Yates (1940)

(*b*) Percentage losses of information for various values of w/w'

w/w'	1	2	4	6	8	12
Case 1	2·21	3·07	4·54	4·37	3·91	—
2	1·73	3·00	3·73	3·19	—	—
3	1·71	2·68*	2·54*	—	—	—
4	2·52	4·04	4·02	3·14	2·53	1·81

* These values are approximate only, being calculated on the assumption that the 24 degrees of freedom for blocks are homogeneous, with mean square expectation $2A+B$.

The actual loss of efficiency depends not only on the numbers of degrees of freedom for M'' and M, but also on the efficiency factor. From the cases already investigated, however, it may be concluded that this source of loss is of little importance in cases likely to occur in practice.

5. Modification when groups of blocks form complete replications

In certain cases the structure of the design is such that blocks fall into groups containing one or more complete replications of all the varieties. When this is so it is clearly advisable in agricultural trials to arrange such groups of blocks in compact large blocks on the ground, since the variation affecting the inter-block varietal estimates will thereby be reduced. Allowance for this must be made in the analysis of variance given above by eliminating complete replications (or groups of replications) from the remainder component of blocks. If there are c such large blocks (containing r/c replications each), the expectations given in Table 2 will require modification as in Table 4.

Table 4. *Expectations of mean squares when groups of blocks contain complete replications*

	D.F.	Expectation
Groups of blocks	$c-1$	—
Varietal component	$v-1$	$EkA+B$
Remainder	$b-v-c+1$	$kA+B$
Varietal component + remainder	$b-c$	$\frac{bk-v-k(c-1)}{b-c}A+B$

The formula for w' will also require modification, being in fact

$$w' = \frac{v(r-1)-k(c-1)}{k(b-c)\,M''-(v-k)\,M}.$$

In the common case in which each large block contains a single replication, $c=r$, and the expectation of the mean square for the $b-r$ degrees of freedom is $\frac{k(r-1)}{r}A+B$, the formula for w' being

$$w' = \frac{r-1}{rM''-M}.$$

6. Simplification when $b = v$, and when $v = k^2$

When $b = v$ the analysis of variance reduces to the simplified form given in Table 5.

Table 5. *Analysis of variance when $b = v$*

	D.F.	S.S.
Blocks (eliminating varieties)	$v-1$	$\frac{dev^2\,W}{rv(v-k)(k-1)}$
Varieties (ignoring blocks)	$v-1$	$\frac{dev^2\,V}{r}$
Intra-block error	$(k-2)v+1$	†
Total	$kv-1$	$dev^2\,y$

There is little point in tabulating the Q's, though they will provide a general check, as before, if this is desired.

A similar simplification is possible in the series of designs $v = k^2$, $r = k+1$, $b = k(k+1)$ (balanced lattices), where the remaining k degrees of freedom for blocks correspond to the contrasts of complete replications.

7. First example

An example of a dummy trial of nine treatments (e.g. dietary treatments) superimposed on the scores of eighteen litters of four rats in a discrimination test is given by Fisher & Yates (1938). Here $v = 9$, $r = 8$, $k = 4$, $b = 18$, $\lambda = 3$, $E = 27/32$.

The individual scores have been given in the publication referred to. Table 6 shows the values of V, T, $4Q$ and W for the nine treatments a–i. The analysis of variance is shown in Table 7, which corresponds in arrangement to Table 1.

Table 6. *Calculation of adjusted scores in discrimination test*

	V	T	$4Q$ $=4V-T$	W $=5V-8T+3G$	Y $=V+\mu W$
a	43·9	152·2	+23·4	+ 65·1	45·7
b	39·1	156·4	0	+ 7·5	39·3
c	41·3	169·6	− 4·4	− 87·1	38·9
d	43·6	151·7	+22·7	+ 67·6	45·4
e	41·7	159·2	+ 7·6	− 1·9	41·6
f	35·6	162·0	−19·6	− 54·8	34·1
g	28·6	138·3	−23·9	+ 99·8	31·3
h	42·8	172·5	− 1·3	−102·8	40·0
i	37·8	155·7	− 4·5	+ 6·6	38·0
	354·4	1417·6	0	0	354·3
Divisor	8	4.5 = 20	$4^2.8.27/32$ = 108	8.9.5.3 = 1080	

Table 7. *Analysis of variance, discrimination test*

	D.F.	S.S. (a)	S.S. (b)	M.S. (b)
Blocks:				
Varietal component	8	41·4684	37·0634	4·6329
Remainder	9	138·2011	138·2011	15·3557
Total	17	179·6695	175·2645	10·3097
Varieties	8	19·6044	24·0094	—
Error	46	119·4506	119·4506	2·5968
Total	71	318·7245	318·7245	—

From the results of the analysis of variance we obtain

$$w = \frac{1}{2\cdot5968} = 0\cdot3851, \quad w' = \frac{63}{68 \times 10\cdot3097 - 5 \times 2\cdot5968} = 0\cdot0916,$$

$$\mu = \frac{0\cdot3851 - 0\cdot0916}{27 \times 0\cdot3851 + 5 \times 0\cdot0916} = \frac{0\cdot2935}{10\cdot8557} = 0\cdot02704.$$

The final adjusted scores in terms of the total scores of eight rats are given in the last column of Table 6. The standard error of these scores is

$$\sqrt{\frac{256}{10\cdot8557}} = \sqrt{23\cdot58} = 4\cdot86.$$

The standard error of the completely adjusted scores (which are equal to Q/E) is

$$\sqrt{(8 \times 2\cdot5968 \times 27/32)} = \sqrt{24\cdot62} = 4\cdot96.$$

Thus the gain in information from the recovery of the inter-block information is $24\cdot62/23\cdot58 - 1$ or 4·4 % (excluding losses due to inaccuracy of weighting). If inter-litter and intra-litter comparisons had been of equal accuracy, the gain would have been 18·5 %.

8. Second example

Table 8 gives the arrangement and yields of a tomato trial of 21 varieties arranged in twenty-one blocks of five plots. (I am indebted to the Statistical Department of Iowa State College for the data of this example.)

Table 8. *Arrangement and yields of a tomato variety trial*

Block ...	1		2		3		4		5		6		7	
	s	22·25	*m*	32·00	*e*	51·75	*r*	45·75	*g*	49·25	*j*	59·00	*b*	61·25
	b	51·50	*l*	44·00	*i*	58·50	*c*	37·25	*a*	33·75	*u*	72·50	*p*	47·75
	o	41·56	*j*	52·50	*k*	29·75	*k*	17·50	*f*	45·75	*o*	49·50	*l*	45·00
	g	36·75	*a*	50·75	*n*	70·75	*q*	26·75	*h*	55·25	*e*	46·50	*t*	35·00
	k	21·00	*k*	32·25	*t*	56·00	*h*	37·25	*i*	62·80	*h*	78·00	*h*	53·00
Total		173·06		211·50		266·75		164·50		246·80		305·50		242·00

Block ...	8		9		10		11		12		13		14	
	e	31·25	*u*	51·00	*k*	24·75	*n*	55·25	*d*	36·50	*o*	38·75	*s*	28·00
	c	35·25	*r*	49·00	*f*	47·25	*o*	37·75	*t*	43·50	*t*	42·25	*q*	40·50
	a	40·50	*a*	40·50	*u*	50·50	*a*	39·50	*q*	35·25	*c*	42·50	*l*	50·25
	b	58·50	*t*	47·75	*p*	58·75	*q*	46·75	*g*	44·00	*f*	50·25	*f*	62·50
	d	45·50	*s*	38·50	*d*	51·25	*p*	48·25	*j*	51·75	*m*	30·75	*e*	41·00
Total		211·00		226·75		232·50		227·50		211·00		204·50		222·25

Block ...	15		16		17		18		19		20		21	
	i	65·00	*n*	57·50	*g*	67·00	*f*	74·00	*b*	67·00	*m*	44·75	*j*	74·25
	d	47·50	*g*	55·00	*r*	70·50	*b*	68·25	*u*	56·75	*d*	62·00	*p*	68·50
	o	49·75	*u*	55·50	*m*	46·00	*r*	86·00	*i*	66·00	*n*	76·75	*c*	46·25
	l	51·00	*c*	38·75	*e*	43·00	*n*	93·25	*m*	31·75	*s*	46·75	*s*	50·25
	r	64·50	*l*	51·25	*p*	64·50	*j*	98·12	*q*	49·00	*h*	82·25	*i*	65·50
Total		277·75		258·00		291·00		419·62		270·50		312·50		304·75

Here $v = 21$, $r = 5$, $k = 5$, $b = 21$, $\lambda = 1$, $E = 21/25$. The values of V, T, W, and the adjusted yields are shown in Table 9, and the analysis of variance in Table 10.*

* Labour would have been saved had the yields been rounded off to 1 decimal place before analysis. The fact that three, and only three, of the yields are not exact quarters may also point to the existence of certain errors of transcription.

Table 9. *Calculation of the adjusted yields, tomato trial*

	V	T	$W = 16V - 20T + 4G$	$Y = V + \mu W$
a	205·00	1123·55	+1927·92	225·57
b	306·50	1316·18	− 300·68	303·29
c	200·00	1142·75	+1463·92	215·62
d	242·75	1244·75	+ 107·92	243·90
e	213·50	1296·50	−1395·08	198·62
f	279·75	1325·67	− 918·48	269·95
g	252·00	1179·86	+1553·72	268·58
h	305·75	1271·30	+ 584·92	311·99
i	317·80	1366·55	−1127·28	305·77
j	335·62	1452·37	−2558·56	308·32
k	125·25	1048·31	+2156·72	148·26
l	241·50	1211·50	+ 752·92	249·53
m	185·25	1290·00	−1717·08	166·93
n	353·50	1484·37	−2912·48	322·43
o	217·31	1188·31	+ 829·68	226·17
p	287·75	1297·75	− 232·08	285·27
q	198·25	1095·75	+2375·92	223·60
r	315·75	1379·62	−1421·48	300·59
s	185·75	1239·31	− 695·28	178·33
t	224·50	1151·00	+1690·92	242·54
u	286·25	1293·25	− 166·08	284·48
	5279·73	26398·65	0	5279·74
Divisor		5	5·21 . 16·4 = 6720	

Table 10. *Analysis of variance, tomato trial*

	D.F.	S.S.	M.S.
Blocks (eliminating varieties)	20	7105·99	355·30
Varieties (ignoring blocks)	20	14222·31	—
Error	64	2363·20	36·92
Total	104	23691·50	—

From Table 10 we have

$$w = 0{\cdot}02709, \quad w' = \frac{84}{100 \times 355{\cdot}30 - 16 \times 36{\cdot}92} = 0{\cdot}00240,$$

$$\mu = \frac{w - w'}{84w + 16w} = \frac{0{\cdot}02469}{2{\cdot}314} = 0{\cdot}01067.$$

The standard error of the adjusted yields is $\sqrt{500/2{\cdot}314} = \sqrt{216{\cdot}1} = 14{\cdot}70$. The standard error of the fully adjusted yields would be $\sqrt{219{\cdot}8}$, so that the gain in information from the use of the inter-block information is trivial. If the inter-block and intra-block comparisons were of equal accuracy the gain would be 19·1 %, less losses due to inaccuracies of weighting.

9. General remarks

In both the examples given the gain in information due to the recovery of the inter-block information is small. Cases will arise, however, in which the chosen blocks do not account for much of the general variability, and in such cases the recovery of the inter-block information will lead to an appreciable increase in efficiency. Since this recovery involves little additional work, and the resulting gain cannot in any case be assessed until the analysis of variance (on the lines set out in this paper) is performed, it would appear best to follow this method of analysis in all cases.

In agricultural experiments, however, the gains from the use of inter-block information will not in general be so great as in similar quasi-factorial (lattice) designs, since complete replications cannot (except in special cases) be arranged in compact groups of blocks. For this reason also, cases will arise in which the use of ordinary randomized blocks will be more efficient than the use of incomplete blocks, whereas lattice designs can never be less efficient than ordinary randomized blocks. Nor is it at all easy, except in data from uniformity trials, to determine exactly what is the efficiency of an incomplete block design, relative to an arrangement in ordinary randomized blocks on the same land.

It will be remembered that lattice designs can be analysed as if they were arrangements in ordinary randomized blocks, the errors of the unadjusted yields being correctly estimated by this process. This property does not hold for incomplete block designs (except those which can be arranged in complete replications) and the full analysis must therefore always be performed.

For these reasons incomplete block designs which cannot be arranged in complete replications are likely to be of less value in agriculture than ordinary lattice designs. Their greatest use is likely to be found in dealing with experimental material in which the block size is definitely determined by the nature of the material. A further use is in co-operative experiments in which each centre can only undertake a limited number of treatments. Here the use of balanced incomplete blocks (each centre forming a block) is frequently much preferable to the common practice of assigning a standard treatment (or control) to each centre.

10. Summary

The recovery of inter-block information in incomplete block designs is discussed, and the method of computation is illustrated by examples.

REFERENCES

R. A. Fisher & F. Yates (1938). *Statistical Tables for Biological, Agricultural and Medical Research.* Edinburgh: Oliver and Boyd.

F. Yates (1936*a*). "Incomplete randomized blocks." *Ann. Eugen., Lond.*, **7**, 121–40.

—— (1936*b*). "A new method of arranging variety trials involving a large number of varieties." *J. Agric. Sci.* **26**, 424–55.

—— (1939). "The recovery of inter-block information in variety trials arranged in three dimensional lattices." *Ann. Eugen., Lond.*, **9**, 136–56.

—— (1940). "Lattice squares." *J. Agric. Sci.* **30**, 672–87.

PAPER IX

ANALYSIS OF DATA FROM ALL POSSIBLE RECIPROCAL CROSSES BETWEEN A SET OF PARENTAL LINES

FROM HEREDITY
VOLUME I, PART 3, pp. 287–301, DECEMBER 1947

Author's Note

Paper IX discusses the analysis of plant-breeding experiments in which all possible reciprocal crosses (diallel crosses) between a group of lines are made and the progeny compared. It provides an interesting example of the application of least squares techniques to experimental material which does not conform to the ordinary factorial structure.

1969 FRANK YATES

ANALYSIS OF DATA FROM ALL POSSIBLE RECIPROCAL CROSSES BETWEEN A SET OF PARENTAL LINES

F. YATES

Rothamsted Experimental Station

I. INTRODUCTION

WHEN assessing the value of different parental lines in plant breeding work on plants which are normally cross-fertilised, the device of making all possible reciprocal crosses and comparing the progeny of these crosses is often of value.

The analysis of the data obtained in such a set of reciprocal crosses can be carried out by an adaptation of the analysis of variance technique. This adaptation, analogous to those appropriate to the

TABLE I

Fertility (number of seeds per 100 florets pollinated) in reciprocal crosses of 12 sibs of an F_1 family of Trifolium hybridum

♀ Parents

♂ Parents	Sib	2	3	7	11	1	5	8	9	10	12	4	6	Total*	Mean*
	2	11	24	74	10	167	134	153	144	87	68	320	155	1228	154
	3	10	24	75	13	95	178	190	163	84	113	61	158	1042	130
	7	9	48	27	33	235	203	158	132	136	252	204	129	1449	181
	11	90	30	58	20	108	170	168	199	119	78	139	150	1131	141
	1	95	112	124	122	21	35	29	20	20	105	223	148	929	133
	5	83	123	123	205	63	50	33	55	21	219	152	148	1053	150
	8	80	114	163	156	41	28	6	27	61	212	130	131	986	141
	9	73	152	162	210	58	32	39	42	64	248	171	237	1253	179
	10	70	125	145	89	61	71	30	69	53	177	143	134	883	126
	12	175	105	218	165	78	266	204	181	85	0	3	123	1600	160
	4	192	116	202	163	150	195	157	154	133	0	2	155	1617	162
	6	120	121	155	178	154	202	163	156	110	170	239	0	1768	161
	Total *	888	968	1292	1288	987	1348	1193	1129	754	1642	1782	1668	14939	
	Mean *	111	121	162	161	141	193	170	161	108	164	178	152		152

* Excluding incompatible crosses.

more complicated forms of experimental design, such as incomplete randomised blocks, was developed to deal with the set of data shown in table I, which gives the fertility in reciprocal crosses of 12 sibs of a certain F_1 family of *Trifolium hybridum* (Alsike). These data were obtained in the course of work at the Welsh Plant Breeding Station, and my thanks are due to Mr Watkin Williams for permission to use them to illustrate the method of analysis.

This particular set of crosses is complicated by the fact that, in addition to nearly complete self-sterility, certain groups of sibs (2, 3, 7, 11), (1, 5, 8, 9, 10), and (12, 4) are incompatible. This incompatibility is clearly shown by the data.

Before discussing the analysis of these results, we will consider the simpler case in which partial or complete self-sterility exists but in which there are no incompatible groups, and also the case in which the selfed plants behave similarly to the crosses.

2. SELF-STERILITY WITHOUT GROUP INCOMPATIBILITY

If the plants are self-sterile, but all crosses are fertile, a table of results of the type given in table 2 will be obtained.

TABLE 2

Data from reciprocal crosses when there is self-sterility

Male parent	Female parent 1	2	3	4	...	Total
1	—	y_{12}	y_{13}	y_{14}	...	$Y_{1\cdot}$
2	y_{21}	—	y_{23}	y_{24}	...	$Y_{2\cdot}$
3	y_{31}	y_{32}	—	y_{34}	...	$Y_{3\cdot}$
4	y_{41}	y_{42}	y_{43}	—	...	$Y_{4\cdot}$
...	...	...	...	...	...	...
Total	$Y_{\cdot 1}$	$Y_{\cdot 2}$	$Y_{\cdot 3}$	$Y_{\cdot 4}$	...	$Y_{\cdot\cdot}$

The y's in this table may represent measurements on single plants, or the means or totals of measurements on a number of plants, possibly derived from a number of replicated plots. The Y's denote the marginal totals of the y's.

It is apparent that the direct comparison of $Y_{1\cdot}$, $Y_{2\cdot}$, . . . will not give valid estimates of the differences between the male parents, since one female parent is missing from each total. $Y_{1\cdot}$, for example, does not contain progeny from female parent 1.

Efficient estimates can be obtained by the method of least squares, fitting constants for the effects of the male and female parents. (See, for example, Yates (1933).) Let there be k parents of each sex, and let the constants be :

Mean : m.
Male parents : $a_1, a_2, a_3, \ldots$ $S(a) = 0$.
Female parents : $b_1, b_2, b_3, \ldots$ $S(b) = 0$.
We then have

$$y_{12} = m + a_1 + b_2 + \epsilon_{12}, \text{ etc.,}$$

where ϵ_{12}, etc., are residuals the sum of whose squares is to be minimised.

The least square equations are as follows :—

$$k(k-1)m+(k-1)\ (a_1+a_2+\ .\ .\ .)+(k-1)\ (b_1+b_2+\ .\ .\ .)=Y_{..}$$
$$(k-1)\ (m+a_1)+b_2+b_3+\ .\ .\ . \qquad =Y_{1\cdot}$$
$$.\quad .\quad .\quad .\quad .\quad .\quad .\quad .$$
$$(k-1)\ (m+b_1)+a_2+a_3+\ .\ .\ . \qquad =Y_{\cdot 1}$$
$$.\quad .\quad .\quad .\quad .\quad .\quad .\quad .$$

From these equations, using the identities $S(a)=0$, $S(b)=0$, we have :

$$m=\bar{y}$$
$$a_1 = \frac{(k-1)\ Y_{1\cdot}+Y_{\cdot 1}-Y_{..}}{k(k-2)} \text{ etc.}$$
$$b_1 = \frac{(k-1)\ Y_{\cdot 1}+Y_{1\cdot}-Y_{..}}{k(k-2)} \text{ etc.}$$

The analysis of variance is shown in table 3, $\text{dev}^2 y$ being used to denote the sum of squares of deviations of all the y's from their mean.

TABLE 3

Analysis of variance of the data of table 2

	Degrees of freedom	Sum of squares
Differences between parents (male and female)	$2(k-1)$	$a_1 Y_{1\cdot}+a_2 Y_{2\cdot}+\ldots+b_1 Y_{\cdot 1}+b_2 Y_{\cdot 2}+\ldots$
Remainder	k^2-3k+1	By difference
Total	$k(k-1)-1$	$\text{dev}^2 y$

This form of analysis is based on the assumption that the effects of the parents, male and female independently, are additive, and that any departures from this additive law are random and independent. If the y's are themselves each derived from a number of replicated plots the remainder term may be compared with the experimental error obtained from an ordinary analysis of these replicates. This comparison will indicate whether there is any departure from the additive law.

The above analysis may be extended in two ways. In the first place, since the effects of the male and female parents of the same line are likely to be similar, we may recast the analysis so as to estimate the average effects of a parent of a particular line (male or female), and the differential effects of the male and female parents of that line. There may, for example, be evidence that the progeny of line 1 are superior to those of other lines, but no evidence that this superiority is more marked for the male or female parent of this line.

In the second place, there may well be some similarity between both reciprocal crosses of a pair of parents, either due to some degree of incompatibility or to some specially favourable conjunction of genes. In this event the y's which are diagonally opposite one another, y_{12} and y_{21}, etc., will be more closely correlated than the y's not so related. This point may be investigated by subdividing the remainder degrees of freedom in the above analysis into those derived from contrasts of the sums of pairs of diagonally opposite values, and those derived from contrasts of the differences of such pairs.

Since the estimates of the combined effects of male and female parents of each line are derived solely from the sums of diagonal pairs, and the differences between male and female parents solely from the differences of diagonal pairs, the whole analysis splits into two parts, one based on the sums, and the other on the differences.

The sums and differences of the diagonal pairs can be set out in two tables of the form given in table 4.

TABLE 4

Sums and differences of diagonal pairs of the data of table 2

(*a*) Diagonal sums, $y_{12}+y_{21}$, etc.

Male parent	Female parent 1	2	3	4	...	Total
1	—	u_{12}	u_{13}	u_{14}	...	U_1
2		—	u_{23}	u_{24}	...	U_2
3			—	u_{34}	...	U_3
4				—	...	U_4
...					...	...
Total						S (U)

(*b*) Diagonal differences, $y_{12}-y_{21}$, etc.

Male parent	Female parent 1	2	3	4	...	Total
1	—	v_{12}	v_{13}	v_{14}	...	V_1
2		—	v_{23}	v_{24}	...	V_2
3			—	v_{34}	...	V_3
4				—	...	V_4
...					...	...
Total						0

The totals U_1, etc., shown in the table of diagonal sums are obtained by summing both the row and column of the parent concerned. Thus $U_3=u_{13}+u_{23}+u_{34}+\ldots$ The totals V_1, etc., in the table of diagonal differences are obtained similarly, except that the values in the column concerned are subtracted. Thus $V_3=-v_{13}-v_{23}+v_{34}+\ldots$ We see that $U_1=Y_{1.}+Y_{.1}$, etc., $V_1=Y_{1.}-Y_{.1}$, etc., and $S(U)=2Y_{..}$

As before, the effects of the parents may be represented by constants, whose values can be estimated by the method of least squares. The two tables can be treated independently, thus providing separate estimates of the residual variances. It will be most convenient to use constants which represent the values of table 4 directly. Let these constants be :

Table 4 (*a*) : Mean : m'
Parents : $c_1, c_2, c_3 \ldots$ $S(c)=0$
Table 4 (*b*) : Differences (male-female) : $d_1, d_2, d_3, \ldots$ $S(d)=0$

Thus

$$u_{12}=m'+c_1+c_2+\epsilon'_{12}$$
$$v_{12}=d_1-d_2+\epsilon''_{12}$$

The least square equations are :

$$\tfrac{1}{2}k(k-1)m'+(k-1)\ (c_1+c_2+\ \ldots)=\tfrac{1}{2}S(U)$$
$$(k-1)\ (m'+c_1)+c_2+c_3+\ \ldots\ =U_1$$
$$\cdot \quad \cdot \quad \cdot \quad \cdot \quad \cdot \quad \cdot$$
$$(k-1)d_1-d_2-d_3\ \ldots\ =V_1$$
$$\cdot \quad \cdot \quad \cdot \quad \cdot \quad \cdot \quad \cdot$$

Whence

$$m'=2\bar{y}$$
$$c_1=\frac{kU_1-2Y..}{k(k-2)}\ \text{etc.}$$
$$d_1=\frac{1}{k}V_1\ \text{etc.}$$

Thus, as we should expect, $c_1=a_1+b_1$ and $d_1=a_1-b_1$, etc.

It will be noted that the difference between the effects of different male parents is given by the differences of the quantities $\tfrac{1}{2}(c+d)$, and of female parents by the differences of $\tfrac{1}{2}(c-d)$. If the differences between male and female parents can be neglected the parental effects (male or female) are measured by the quantities $\tfrac{1}{2}c$. Thus, under these circumstances, the expected difference between the cross with parents 1 and 2, and the cross with parents 1 and 3, is $\tfrac{1}{2}(c_2-c_3)$; that between the crosses with parents 1 and 2, and with parents 3 and 4, is $\tfrac{1}{2}(c_1+c_2)-\tfrac{1}{2}(c_3+c_4)$.

The analysis of variance now splits into two parts, as shown in table 5. The extra factor of $\tfrac{1}{2}$ has been introduced into the sums of squares to bring them to units of a single entry of table 2.

TABLE 5

Analysis of variance of sums and differences of diagonal pairs

		Degrees of freedom	Sum of squares
Diagonal sums (*a*)	Parents	$k-1$	$\frac{1}{2(k-2)}\ \text{dev}^2U$
	Remainder	$\tfrac{1}{2}(k-2)\ (k-1)-1$	By difference
	Total	$\tfrac{1}{2}k(k-1)-1$	$\tfrac{1}{2}\text{dev}^2u$
Diagonal differences (*b*)	Parents	$k-1$	$\frac{1}{2k}\ S(V^2)$
	Remainder	$\tfrac{1}{2}(k-2)\ (k-1)$	By difference
	Total . .	$\tfrac{1}{2}k(k-1)$	$\tfrac{1}{2}S(v^2)$

The contrast between the mean squares for remainder (*a*) and remainder (*b*) will indicate whether there is any significantly greater similarity between reciprocal crosses of the same parents than between crosses of different parents, after allowance for the average effects of parents, male and female independently, has been made. In order to test if there are any differences between parents which are consistent for all crosses the mean square for parents (*a*) should be compared with remainder (*a*). Comparison between parents (*b*) and remainder (*b*) gives the similar test for consistency of differences between male and female parents. If an estimate of experimental error is available the remainder mean squares may be compared with the error mean square.

The standard errors of the *c*'s and *d*'s are given by the formulæ

$$V(c) = \frac{1}{(k-2)}(2\sigma_1^2)$$

$$V(d) = \frac{1}{k}(2\sigma_2^2)$$

the appropriate estimates of σ_1^2 and σ_2^2 being obtained from the analysis of variance, and the factor 2 being introduced because each value of table 4 is the sum or difference of two values in table 2. In general the mean square for remainder (*a*) will provide the appropriate estimate for σ_1^2, and that for remainder (*b*) for σ_2^2, though if experimental errors only are under consideration the mean square for experimental error, if available, will be appropriate.

3. NO SELF-STERILITY

In the absence of self-sterility the data can be arranged in a $k \times k$ table similar to table 2, but with the diagonal cells occupied. The analysis of variance of this table presents no difficulties, since the two sets of marginal totals are orthogonal, and provide estimates of the differences between parents, male and female independently. The corresponding totals may be added to provide estimates of the differences between parents averaged over both sexes, and subtracted to provide estimates of the differential effects of the two sexes.

The possibility of the existence of greater similarity between reciprocal crosses can best be tested by the procedure of section 2, omitting the progeny of selfed parents. The inclusion of the selfed matings in this analysis introduces additional complications, and would add little to the information provided by the partial analysis.

4. SELF-STERILITY WITH MUTUALLY INCOMPATIBLE GROUPS

If certain groups of parents are mutually incompatible the general procedure of section 2 can be followed, though the solution is somewhat more complicated, owing to the fact that all comparisons will not be of the same accuracy. The least square equations, however, are still directly solvable.

Let the k parents be divided into groups of $p, q, r, \ldots$ mutually incompatible parents, so that $p+q+r+\ldots=k$, and denote the constants, etc., of the first group of parents by the suffixes $p1, p2, \ldots$ with the convention that $c_{p1}+c_{p2}+\ldots=\mathrm{S}(c_p)$, etc.

The least square equations are:

Diagonal sums

$$\tfrac{1}{2}\{p(k-p)+q(k-q)+\ldots\}m'+(k-p)\mathrm{S}(c_p)+(k-q)\mathrm{S}(c_q)+\ldots=\tfrac{1}{2}\mathrm{S}(\mathrm{U})$$

Group p (p equations)

$$\begin{cases}(k-p)(m'+c_{p1})+\mathrm{S}(c_q)+\mathrm{S}(c_r)+\ldots=\mathrm{U}_{p1}\\ \cdot\quad\cdot\quad\cdot\quad\cdot\quad\cdot\quad\cdot\quad\cdot\end{cases}$$

Group q (q equations)

$$\begin{cases}(k-q)(m'+c_{q1})+\mathrm{S}(c_p)+\mathrm{S}(c_r)+\ldots=\mathrm{U}_{q1}\\ \cdot\quad\cdot\quad\cdot\quad\cdot\quad\cdot\quad\cdot\quad\cdot\end{cases}$$

. . . .

Diagonal differences

Group p (p equations)

$$\begin{cases}(k-p)d_{p1}-\mathrm{S}(d_q)-\mathrm{S}(d_r)-\ldots=\mathrm{V}_{p1}\\ \cdot\quad\cdot\quad\cdot\quad\cdot\quad\cdot\quad\cdot\end{cases}$$

Group q (q equations)

$$\begin{cases}(k-q)d_{q1}-\mathrm{S}(d_p)-\mathrm{S}(d_r)-\ldots=\mathrm{V}_{q1}\\ \cdot\quad\cdot\quad\cdot\quad\cdot\quad\cdot\quad\cdot\quad\cdot\end{cases}$$

. . . .

The c and d equations may be rewritten

$$\left.\begin{array}{r}(k-p)(m'+c_{p1})-\mathrm{S}(c_p)=\mathrm{U}_{p1}\\ \cdot\quad\cdot\quad\cdot\quad\cdot\quad\cdot\quad\cdot\\ (k-q)(m'+c_{q1})-\mathrm{S}(c_q)=\mathrm{U}_{q1}\\ \cdot\quad\cdot\quad\cdot\quad\cdot\quad\cdot\quad\cdot\\ (k-p)d_{p1}+\mathrm{S}(d_p)=\mathrm{V}_{p1}\\ \cdot\quad\cdot\quad\cdot\quad\cdot\quad\cdot\quad\cdot\\ (k-q)d_{q1}+\mathrm{S}(d_q)=\mathrm{V}_{q1}\\ \cdot\quad\cdot\quad\cdot\quad\cdot\quad\cdot\quad\cdot\end{array}\right\}\quad(\mathrm{A})$$

Summing the equations of each group, we have:

$$\begin{array}{r}p(k-p)m'+(k-2p)\mathrm{S}(c_p)=\mathrm{S}(\mathrm{U}_p)\\ q(k-q)m'+(k-2q)\mathrm{S}(c_q)=\mathrm{S}(\mathrm{U}_q)\\ \cdot\quad\cdot\quad\cdot\quad\cdot\quad\cdot\quad\cdot\quad\cdot\quad\cdot\\ k\mathrm{S}(d_p)=\mathrm{S}(\mathrm{V}_p)\\ k\mathrm{S}(d_q)=\mathrm{S}(\mathrm{V}_q)\\ \cdot\quad\cdot\quad\cdot\quad\cdot\end{array}$$

Hence

$$\mathrm{S}(c_p)=\frac{1}{k-2p}\mathrm{S}(\mathrm{U}_p)-\frac{p(k-p)}{k-2p}m'$$

etc.

Since $S(c_p)+S(c_q)+\ldots = 0$ we have

$$Km' = \frac{1}{k-2p}S(U_p)+\frac{1}{k-2q}S(U_q)+\ldots$$

where

$$K = \frac{p(k-p)}{k-2p}+\frac{q(k-q)}{k-2q}+\ldots$$

It may be noted that since $p+q+\ldots = k$

$$K = \tfrac{1}{2}k\left\{1+\frac{p}{k-2p}+\frac{q}{k-2q}+\ldots\right\}$$

The above equation gives the numerical value of m' (which will not in this case be exactly equal to twice the mean of the original values). By substitution for $S(c_p)$ and $S(d_p)$ in equations (A) we also find

$$(k-p)c_{p1} = U_{p1}+\frac{1}{k-2p}\{S(U_p)-(k-p)^2m'\}$$

. . . .

$$(k-p)d_{p1} = V_{p1}-\frac{1}{k}S(V_p)$$

. . . .

It will be noted that the first of the original least square equations has not been used. This equation is redundant in virtue of the identity $S(c) = 0$.

The degrees of freedom and sums of squares in the analysis of variance are given in table 6, in which $N = \tfrac{1}{2}\{p(k-p)+q(k-q)+\ldots\}$, the number of entries in the table of diagonal sums and differences.

TABLE 6

Analysis of variance with self-sterility and mutually incompatible groups

		Degrees of freedom	Sum of squares
Diagonal sums (*a*)	Parents	$k-1$	$\tfrac{1}{2}\{\tfrac{1}{2}m'S(U)+c_{p1}U_{p1}+\ldots+c_{q1}U_{q1}+\ldots\}-\dfrac{\{\tfrac{1}{2}S(U)\}^2}{2N}$
	Remainder	$N-k$	By difference
	Total	$N-1$	$\tfrac{1}{2}\text{dev}^2u$
Diagonal differences (*b*)	Parents	$k-1$	$\tfrac{1}{2}(d_{p1}V_{p1}+\ldots+d_{q1}V_{q1}+\ldots)$
	Remainder	$N-k+1$	By difference
	Total . .	N	$\tfrac{1}{2}S(v^2)$

It will be noted that, since m' is not equal to the mean, it must be introduced explicitly in the sum of squares for the constants, and the ordinary correction for the mean deducted.

The formulæ for the standard errors of the differences of the c's and d's may be derived from the solution of auxiliary sets of equations in the manner followed in partial regression analysis. There are certain points of difference, however, which are of general recurrence in least square solutions of this type, and which will therefore be worth describing. For this description we will use the notation customarily adopted in regression analysis (with the exception that, to avoid confusion, the c's of this notation will be replaced by c''s). In this notation the regression coefficients b_1, b_2, ..., satisfy the equations

$$b_1 S(x_1^2) + b_2 S(x_1 x_2) + \dots = S(x_1 y)$$
$$b_1 S(x_1 x_2) + b_2 S(x_2^2) + \dots = S(x_2 y)$$
$$\cdot \quad \cdot \quad \cdot \quad \cdot \quad \cdot \quad \cdot \quad \cdot$$

c'_{11}, c'_{12}, c'_{13}, ... satisfy a similar set of equations with the numerical terms $S(x_1 y)$, $S(x_2 y)$, $S(x_3 y)$, ... replaced by 1, 0, 0, ... , c'_{21}, c'_{22}, c'_{23} a similar set with numerical terms 0, 1, 0, ... , etc., and $c'_{12} = c'_{21}$ etc. Also

$$V(b_1) = c'_{11}\sigma^2,\ V(b_2) = c'_{22}\sigma^2, \text{ etc.,}$$
$$\text{cov}(b_1 b_2) = c'_{12}\sigma^2, \text{ etc.,}$$

so that

$$V(b_1 - b_2) = (c'_{11} + c'_{22} - 2c'_{12})\sigma^2.$$

In regression analysis the numerical solution of the equations for the b's is usually effected by first determining the matrix of c''s, and thence the b's from the relations

$$b_1 = c'_{11} S(x_1 y) + c'_{12} S(x_2 y) + \dots$$
$$\cdot \quad \cdot \quad \cdot \quad \cdot \quad \cdot \quad \cdot \quad \cdot$$

In cases such as the present, in which the equations are such that a simple algebraic solution is possible, the procedure may be reversed, and the c''s determined from this solution.

If there are redundant regression coefficients or constants such that

$$\mu_1 b_1 + \mu_2 b_2 + \mu_3 b_3 + \dots = \mu_0$$
$$\mu'_1 b_1 + \mu'_2 b_2 + \mu'_3 b_3 + \dots = \mu'_0$$
$$\cdot \quad \cdot \quad \cdot \quad \cdot \quad \cdot \quad \cdot \quad \cdot$$

it has been shown (Yates and Hale, 1939) that the numerical terms 1, 0, 0, . . . of the set of equations for c'_{11}, c'_{12}, c'_{13}, . . . must be replaced by

$$1 - \mu_1 d - \mu'_1 d' - \dots,\ -\mu_2 d - \mu'_2 d' - \dots,\ -\mu_3 d - \mu'_3 d' - \dots,\ \dots$$

where d, d', . . . are so chosen that the relations to which the redundant constants give rise between the coefficients of the normal equations still persist.

In the present case we have, in the solution for the diagonal sums,

$$c_{p1}+c_{p2}+\ \dots\ +c_{q1}+c_{q2}+\ \dots\ +c_{r1}+c_{r2}+\ \dots\ =0$$

and the μ's for all the constants except m are therefore unity. Also the sum of the numerical terms of all the equations except the first, k in number, is equal to twice the numerical term of the first equation. Consequently the numerical terms for the set of equations giving $c'_{p1\cdot p1}$, etc., are

$$0,\ 1-\frac{1}{k},\ -\frac{1}{k},\ -\frac{1}{k},\ \dots$$

Hence the values for $c'_{p1\cdot p1}$, etc., are given by the substitutions

$$U_{p1}=1-\frac{1}{k},\quad U_{p2}=U_{p3}=\dots=U_{q1}=\dots=-\frac{1}{k}$$

Substituting these values in the solutions already obtained and denoting the resultant value of m' by m'_{p1}, we find

$$m'_{p1}=\frac{1}{kK}\left\{\frac{k-p}{k-2p}-\frac{q}{k-2q}-\frac{r}{k-2r}-\dots\right\}$$

$$=\frac{2}{kK}\,\frac{k-p}{k-2p}-\frac{2}{k^2}$$

$$c'_{p1\cdot p1}=\frac{1}{k-p}\left[1-\frac{1}{k}+\frac{1}{k-2p}\left\{\frac{k-p}{k}-(k-p)^2\,m'_{p1}\right\}\right]$$

$$c'_{p1\cdot p2}=\frac{1}{k-p}\left[-\frac{1}{k}+\frac{1}{k-2p}\left\{\frac{k-p}{k}-(k-p)^2\,m'_{p1}\right\}\right]$$

$$c'_{p1\cdot q1}=\frac{1}{k-q}\left[-\frac{1}{k}+\frac{1}{k-2q}\left\{-\frac{q}{k}-(k-q)^2\,m'_{p1}\right\}\right]$$

The other c''s can be written down immediately from these expressions by symmetry, and it can be verified that $c'_{p1\cdot q1}=c'_{q1\cdot p1}$.

We now have

$$V(c_{p1}-c_{p2})=(c'_{p1\cdot p1}+c'_{p2\cdot p2}-2c'_{p1\cdot p2})\sigma^2$$

$$=\frac{2}{k-p}\,(2\sigma_1^2)$$

$$V(c_{p1}-c_{q1})=(c'_{p1\cdot p1}+c'_{q1\cdot q1}-c'_{p1\cdot q1}-c'_{q1\cdot p1})\sigma^2$$

$$=(2\sigma_1^2)\left\{\frac{1}{k-p}+\frac{1}{k-q}+\frac{1}{(k-p)(k-2p)}+\frac{1}{(k-q)(k-2q)}-\frac{2k(p-q)^2}{K(k-2p)^2(k-2q)^2}\right\}$$

where, as before, estimates of σ_1^2 are given by the remainder mean squares of the analysis of variance (table 6). These two expressions give the variances of the difference of any pair of parents in the same or in different groups.

The same procedure may be followed for the diagonal differences. It will be found that

$$V(d_{p1}-d_{p2}) = \frac{2}{k-p}(2\sigma_2^2)$$

$$V(d_{p1}-d_{q1}) = \left(1-\frac{1}{k}\right)\left(\frac{1}{k-p}+\frac{1}{k-q}\right)(2\sigma_2^2)$$

It can be verified that when $p = q = 1$ the expressions for $V(c_{p1}-c_{q1})$ and $V(d_{p1}-d_{q1})$ reduce to the expressions already found for the case in which there is no group incompatibility.

5. ANALYSIS OF NUMERICAL DATA

We may now consider the analysis of the data of table 1. In this case $k = 12$, $p = 4$, $q = 5$, $r = 2$, $s = 1$, the parents in the four groups being sibs (2, 3, 7, 11), (1, 5, 8, 9, 10), (12, 4), and (6) respectively.

The sums and differences of the diagonal pairs are shown in table 7*a* and table 7*b*, the incompatible crosses being omitted. The U and V totals, and $S(U_p)$, and $S(V_p)$, etc., are also shown in these tables.

TABLE 7*a*

Diagonal sums of data of table 1

♂ Parent	♀ Parent 1	5	8	9	10	12	4	6	U	S(U)
2	262	217	233	217	157	243	512	275	2116	
3	207	301	304	315	209	218	177	279	2010	9286
7	359	326	321	294	281	470	406	284	2741	
11	230	375	324	409	208	243	302	328	2419	
1						183	373	302	1916	
5						485	347	350	2401	
8						416	287	294	2179	10515
9						429	325	393	2382	
10						262	276	244	1637	
12								293	3242	6641
4								394	3399	
6									3436	3436
										29878

TABLE 7*b*

Diagonal differences of data of table 1

♂ Parent	♀ Parent 1	5	8	9	10	12	4	6	V	S(V)
2	+72	+51	+73	+71	+17	−107	+128	+35	+340	
3	−17	+55	+76	+11	−41	+8	−55	+37	+74	+414
7	+111	+80	−5	−30	−9	+34	+2	−26	+157	
11	−14	−35	+12	−11	+30	−87	−24	−28	−157	
1						+27	+73	−6	−58	
5						−47	−43	−54	−295	
8						+8	−27	−32	−207	−307
9						+67	+17	+81	+124	
10						+92	+10	+24	+129	
12								−47	−42	−207
4								−84	−165	
6									+100	+100
										0

The equations for K, m', the c's and the d's are as follows :—

$$K = 6\left\{1+\frac{4}{4}+\frac{5}{2}+\frac{2}{8}+\frac{1}{10}\right\} = 29{\cdot}1$$

$$m' = \frac{1}{29{\cdot}1}\left\{\frac{1}{4}9286+\frac{1}{2}10515+\frac{1}{8}6641+\frac{1}{10}3436\right\} = 300{\cdot}7809$$

$$c_{p1} = \frac{1}{8}\left[2116+\frac{1}{4}\left\{9286-8^2\times 300{\cdot}7809\right\}\right] = \frac{1}{8}2116-311{\cdot}3743 = -46{\cdot}8743$$

$$c_{p2} = \frac{1}{8}2010-311{\cdot}3743 = -60{\cdot}1243$$

. . . .

$$c_{q1} = \frac{1}{7}\left[1916+\frac{1}{2}\left\{10515-7^2\times 300{\cdot}7809\right\}\right] = \frac{1}{7}1916-301{\cdot}6617 = -27{\cdot}9474$$

. . . .

$$d_{p1} = \frac{1}{8}\left[+340-\frac{1}{12}(+414)\right] = +38{\cdot}1875$$

. . . .

$$d_{q1} = \frac{1}{7}\left[-58-\frac{1}{12}(-307)\right] = -4{\cdot}6310$$

. . . .

The complete set of values for the c's and the d's are given in table 8.

TABLE 8

Values of c's and d's

Group	Parent	c	d
p	2	−46·8743	+38·1875
	3	−60·1243	+4·9375
	7	+31·2507	+15·3125
	11	−8·9993	−23·9375
q	1	−27·9474	−4·6310
	5	+41·3383	−38·4881
	8	+9·6240	−25·9167
	9	+38·6240	+21·3690
	10	−67·8046	+22·0833
r	12	+31·2364	−2·4750
	4	+46·9364	−14·7750
s	6	+12·7410	+8·3333
		+0·0009	−0·0002

The sums of squares for parents in the analysis of variance are calculated as follows :—

$$S(cU) = -46{\cdot}8743 \times 2116 + \ldots \quad = +196119$$

$$\tfrac{1}{2}m'S(U) = \frac{1}{2} \times 300{\cdot}7809 \times 29878 \quad = +4493366$$

$$-\frac{1}{4N}\left\{S(U)\right\}^2 = -\frac{1}{4\times 49} \times 29878^2 = -4554566$$

$$134919$$

$$S(dV) = (+38{\cdot}1875) \times (+340) + \ldots = \quad 45372$$

The sum of squares of deviations of the values in the body of table 7*a*, dev^2u, is 319,721, and the sum of squares of the values of table 7*b*, $S(v^2)$, is 140,559. Introducing the further factor of $\frac{1}{2}$ we obtain the analysis of variance shown in table 9.

The main features of the data are now apparent. The mean square for sibs (a) is significantly greater than the remainder (a) at the 5 per cent. level, and there is also some indication of a difference between sibs (b) and remainder (b), though this is not fully significant. Furthermore, remainder (a) is significantly greater than remainder (b) at the 5 per cent. level ($e^{2z} = 1{\cdot}99$, 5 per cent. value $= 1{\cdot}70$). There is thus clear indication of differences between sibs, which have a certain degree of consistency over all compatible matings. There is some indication of differences in effect between pollen and ova of the same sib, though these differences, if they exist, are not large.

Finally, crosses between individual pairs of sibs deviate somewhat from the additive law based on the average performance of either sib separately in all compatible crosses.

TABLE 9

Analysis of variance of data of table 1

		Degrees of freedom	Sum of squares	Mean square
Diagonal sums (*a*)	Sibs	11	67460	6133
	Remainder	37	92400	2497
	Total	48	159860	
Diagonal differences (*b*)	Sibs	11	22686	2062
	Remainder	38	47594	1252
	Total	49	70280	

The final results may be presented in a table of values of $\frac{1}{2}(c+d)$, $\frac{1}{2}(c-d)$, $\frac{1}{2}c$ and d, giving respectively the estimates of the average effects of the male parents, the female parents, the mean of male and female parents, and the difference of male and female parents. The standard errors of individual comparisons may be calculated, if required, from the formulæ already given. Allowing for irregularities in the performance of the different crosses, *i.e.* basing the error variance on remainder (*a*) of table 9, the smallest of the standard errors for c comparisons, that between sib 12 and sib 4, for example, is given by

$$V(c_{r1}-c_{r2}) = \frac{2}{10}(2497\times 2) = 31\cdot 6^2$$

Similarly the largest of these standard errors, that between any member of group p and any member of group q, is given by

$$V(c_{p1}-c_{q1}) = \left\{\frac{1}{8}+\frac{1}{7}+\frac{1}{8\times 4}+\frac{1}{7\times 2}-\frac{2\times 12(8-7)^2}{29\cdot 1\times 4^2\times 2^2}\right\}(2497\times 2) = 42\cdot 2^2$$

The smallest and largest standard errors for the d comparisons are similarly given by

$$V(d_{r1}-d_{s1}) = \frac{11}{12}\left(\frac{1}{10}+\frac{1}{11}\right)\times 1252\times 2 = 20\cdot 9^2$$

$$V(d_{q1}-d_{q2}) = \frac{2}{7}\times 1252\times 2 = 26\cdot 7^2$$

Since the c and d comparisons are independent the standard errors of the comparisons of $\frac{1}{2}(c+d)$ and $\frac{1}{2}(c-d)$ can be obtained from the above standard errors by the ordinary rules for the combination of variances of independent variates.

For many purposes average standard errors which are approximately applicable to all comparisons will be adequate. These can be calculated directly from the mean square deviations of the values in table 8 and the corresponding variance ratios in table 9. The variance ratio for the c comparisons is 6133/2497 = 2·46, and the mean square deviation of the c values is 18883/11 = 1717. The average standard error of a single c value is therefore approximately $\sqrt{1717/2\cdot46} = \pm 26\cdot4$. The average standard error of the difference of two values is therefore $\sqrt{2}$ times this or $\pm 37\cdot3$. Similarly the average standard error of a single d value is $\sqrt{518/1\cdot65} = \pm 17\cdot7$ and of the difference of two values is $\pm 25\cdot0$.

Examining the results of table 8 in detail, we see that five sibs, 7, 5, 9, 12 and 4, give decidedly better performance than average. The crosses between these sibs have picked out a good proportion of the high values of table 7*a*, but there are some remaining, in particular those between 4 and 2, between 4 and 6, between 9 and 11, between 9 and 6, and between 12 and 8 of which one parent is a sib not in this group. Again, some of the crosses between the sibs of the top group, in particular those between 7 and 5, 7 and 9, and 9 and 4, have given values decidedly below expectation. These inconsistencies are, in part at least, a reflection of departures from the additive law. If it is desired to pick pairs of parents which may be expected to give high fertility, we should in these circumstances give some weight to the performance of the individual crosses as well as the performance of either parent separately. We should also perform the cross in the direction which the d value indicates is most favourable, *e.g.* in a mating between 5 and 12 we should use sib 5 as the female parent.

6. SUMMARY

The paper describes the analysis of data obtained in plant breeding work when all possible reciprocal crosses between different lines are made. The cases discussed are : self-sterility, no self-sterility, self-sterility with incompatibility within groups of lines. The last case is illustrated by a numerical example.

REFERENCES

YATES, F. 1933.
The principles of orthogonality and confounding in replicated experiments.
J. Agric. Sci. *23*, 108-145.

YATES, F., and HALE, R. W. 1939.
The analysis of Latin squares when two or more rows, columns, or treatments are missing.
Suppl. J. R. Statist. Soc. *6*, 67-79.

PAPER X

THE ANALYSIS OF EXPERIMENTS CONTAINING DIFFERENT CROP ROTATIONS

FROM BIOMETRICS
VOLUME 10, pp. 324–346, 1954

Author's Note

Both the design and the analysis of long-term experiments present many problems which do not arise in experiments in which there are no changes in the treatments during the course of the experiment. Experiments which include different crop rotations in the same experiment are particularly troublesome, because only plots having the same crop in a given year can be meaningfully compared. Paper X discusses the analysis of an experiment of this type which contained rotations of different length. This gave rise to additional complexities. The paper is chiefly of interest to those who wish to study in depth the relation between the structure of data and analysis of variance techniques.

1969 FRANK YATES

THE ANALYSIS OF EXPERIMENTS CONTAINING DIFFERENT CROP ROTATIONS

F. YATES
Rothamsted Experimental Station.

Summary

The problems arising in the analysis of experiments containing different crop rotations are investigated. When the design of the experiment is such that each block contains plots which sometimes carry a given crop but do not all carry the crop in the same set of years the year-block totals will not be orthogonal with the plot totals. In most such cases the fitting of constants must be resorted to in order to obtain separate estimates of plot error and plot × year error which are free of year × block interactions. The method is illustrated by application to a rice-pasture experiment containing rotations of different lengths and with different proportions of rice to pasture.

Introduction

The present paper deals only with problems arising in the analysis (not the design) of experiments containing different crop rotations, i.e. experiments of type (b) below, mainly in relation to a proposed experiment on alternative rice-pasture rotations in the United States. Some problems arising in the analysis of experiments comparing different treatments on the same rotation, i.e. experiments of type (a) below, have been discussed by Patterson (1953). A general discussion of the design of rotation experiments has been given by Yates (1949). An earlier discussion of the design and analysis of long-term experiments is provided by Cochran (1939), and some points arising in the analysis of experiments of type (b) are considered by Crowther and Cochran (1942).

Terminology

The design and analysis of experiments involving crop rotations is one that has received relatively little notice in the literature, and a brief note on the terminology of the subject may therefore be helpful to the understanding of this paper.

Sequence. In this paper the term sequence is used to denote a sequence of crops, treatments or crop-treatment combinations which differs in any respect from other such sequences occurring in the experiment.

Cycle, period, phase. These terms are used in their customary sense. If we have a repetitive sequence of crops or treatments c_1, c_2, $\cdots$, c_n, c_1, c_2, $\cdots$, c_n a single repetition is termed a cycle. The number of years (or other time units) in the cycle is termed its period. Sequences starting at different points in the same cycle are said to be in different phase. There are thus n possible phases of a cycle of period n. The term phase is also used to denote the different components of the cycle. These alternative meanings are not likely to cause confusion and follow common usage (e.g. phases of an alternating electric current and phases of the moon).

If two or more cycles of differing period are included in the experiment then the whole experiment will follow a cycle whose period is the lowest common multiple of the two or more periods.

Rotation. A rotation is a definite cycle of crops grown in successive years on the same land. In agricultural practice the term is used somewhat loosely, and includes not only minor variations in cropping from cycle to cycle, e.g. the substitution of one cereal crop for another, but often also alterations in the length of the cycle. In experimental work such variations are naturally kept to a minimum. The separate crops of a rotation are in agricultural terminology called courses. If the same crop occurs twice in a rotation it will constitute two different courses. An n course rotation is therefore a cropping cycle of period n.

Rotation experiments. There are two main classes of rotation experiments:

(a) Experiments on the effects of treatments applied to a fixed rotation of crops. The experimental treatments may be repeated year after year, or may be varied in some manner which is regarded as appropriate to the questions at issue. A common device is to use a cycle of treatments. If the cycle of treatments on a given plot of a rotation experiment has the same period as the cropping cycle a given crop will have the same treatment each time it is grown. If the periods are different the treatment will vary in a cyclic manner, with the important consequence that any given combination of crop and treatment-phase occurs on different plots in successive crop cycles, thereby considerably increasing the accuracy.

(b) Experiments comparing the effects of different rotations. Here the different crops themselves act as treatments, and plots between which comparisons have to be made will not always be carrying the same crop in the same year. This introduces the additional problem of design of ensuring that the plots to be compared simultaneously carry the same crop in a sufficient number of years for the necessary comparisons to be made, a problem which is particularly troublesome when the effect of rotations of different length is under investigation.

Experiments of class (b) may, and often do, contain other experimental treatments in addition to the variations in cropping.

Series. In rotation experiments of type (a) the area of land is usually divided into separate parts for the different crop phases. Except when the same crop occurs more than once in the cycle each part will in any one year carry a different crop. These separate parts are termed series.

Blocks. The whole experiment, or if in series the separate series, will normally be divided into blocks, as in ordinary one-year agricultural experiments. Each block may contain a complete replicate of all the sequences in the experiment or the series, or the device of confounding may be used to reduce block size, in which case each block will only contain part of a replicate.

Preliminary years. One, two or more years at the beginning of a cyclical experiment usually have to be excluded from the main analysis because treatments which would have been applied had the experiment been started earlier will not in fact have been applied. These years are known as the preliminary years. Their number will depend on the nature of the treatments and other factors.

Phase differences. In rotation experiments many different types of contrast may be of interest. One type, which has no analogy in one-year experiments, is the contrast between different phases of the same cycle. Such contrasts may be termed phase differences. In a crop rotation containing the same crop twice in the rotation, for instance, there will be a phase difference for this crop. This phase difference can only be estimated with precision if the phases carrying the same crop occur in the same blocks. Thus with the crop rotation PQPR estimation of the phase difference requires arrangement in two series instead of four. With the rotation PPQR a single series must be used if estimates are to be available each year.

The problem

At the Summer Statistics Conference organized by the North Carolina Institute of Statistics in 1952 and held at Blue Ridge, North Carolina, an experiment to compare various rotations of rice and grass pasture was discussed. Comparisons were required between the rotations:

A. 1 year rice, 2 years grass
B. 2 years rice, 2 years grass
C. 1 year rice, 3 years grass

The design proposed was to compare all phases of these three rotations in three randomised blocks of 11 plots each. A complete cycle of this experiment requires 12 years, after 3 preliminary years. One replicate of such a cycle (excluding the preliminary years) is shown in Table 1.

TABLE 1. ONE REPLICATE OF THE RICE-PASTURE ROTATION EXPERIMENT

Year	A letter denotes the plot is carrying rice.											Total
	A			B				C				
	1	2	3	4	5	6	7	8	9	10	11	
1	a_1			b_1			b_1'	c_1				Y_1
2		a_2		b_2'	b_2				c_2			Y_2
3			a_3		b_3'	b_3				c_3		Y_3
4	a_4					b_4'	b_4				c_4	Y_4
5		a_5		b_5			b_5'	c_5				Y_5
6			a_6	b_6'	b_6				c_6			Y_6
7	a_7				b_7'	b_7				c_7		Y_7
8		a_8				b_8'	b_8				c_8	Y_8
9			a_9	b_9			b_9'	c_9				Y_9
10	a_{10}			b_{10}'	b_{10}				c_{10}			Y_{10}
11		a_{11}			b_{11}'	b_{11}				c_{11}		Y_{11}
12			a_{12}			b_{12}'	b_{12}				c_{12}	Y_{12}
Total	P_1	P_2	P_3	P_4	P_5	P_6	P_7	P_8	P_9	P_{10}	P_{11}	R

The yields of rice in the three rotations can be assessed by comparing the means of the a, $\frac{1}{2}(b + b')$ and c yields, and the difference between the first and second year's rice in the B rotation (which we may term the crop phase difference) can be assessed by comparing the means of the b and b' yields. All these comparisons will be free of year differences since each year is equally represented in each mean. The first group of comparisons will involve plot differences, whereas the phase comparisons will only involve plot and year interaction. Similar comparisons can be made over a shorter period than 12 years, though some of the symmetry is then lost. If only six years results are available, for example, each of the A plots is represented twice, but of the B plots plot 4 is represented four times, plots 5 and 7 three times, and plot 6 twice, and similarly with the C plots.

The problem is to estimate the errors of these and other comparisons that may require to be made. The ordinary subdivision of the analysis of variance into a part derived from the totals of plots over all years, and another part (further sub-divided if necessary) derived from the plots × years interaction, breaks down, since the different plot totals involve year differences.

The full analysis of an experiment of this type provides an interesting example of the partition of degrees of freedom in material which possesses a certain degree of balance and orderliness but is not fully orthogonal. In order to elucidate the various points at issue we will first consider the analysis of two somewhat simpler types of experiment.

A rotation with all crops different, treatments tied to crops

As a simple example we may take a three course rotation experiment to compare two treatment cycles (or fixed treatments), with three replicates of each of the three phases. Denote the treatment cycles by A and B. It is assumed that a given crop always has the same treatment i.e. that the periods of the crop and treatment cycles are the same. Thus we might have an experiment in which treatment cycle A was the application of farmyard manure to the potato crop in a potatoes-barley-wheat rotation, treatment cycle B being the control (no farmyard manure). The barley and wheat crops will then measure the first and second year residual effects of the farmyard manure.

As all the crops are different each phase of the rotation will normally be grouped in a separate series of three blocks. Table 2 shows the 6

TABLE 2. CROP TREATMENT SEQUENCES IN A THREE COURSE ROTATION WITH ALL CROPS DIFFERENT

(Crops: P, Q, R; treatment cycles: A_1, A_2, A_3; B_1, B_2, B_3)

	Series: Blocks:	I 1, 2, 3		II 4, 5, 6		III 7, 8, 9	
	Sequence:	1	2	3	4	5	6
Preliminary years	−1	$Q\ A_2$	$Q\ B_2$	$P\ A_1$	$P\ B_1$	$R\ A_3$	$R\ B_3$
	0	$R\ A_3$	$R\ B_3$	$Q\ A_2$	$Q\ B_2$	$P\ A_1$	$P\ B_1$
Experimental years	1	$P\ A_1$	$P\ B_1$	$R\ A_3$	$R\ B_3$	$Q\ A_2$	$Q\ B_2$
	2	$Q\ A_2$	$Q\ B_2$	$P\ A_1$	$P\ B_1$	$R\ A_3$	$R\ B_3$
	3	$R\ A_3$	$R\ B_3$	$Q\ A_2$	$Q\ B_2$	$P\ A_1$	$P\ B_1$
	4	$P\ A_1$	$P\ B_1$	$R\ A_3$	$R\ B_3$	$Q\ A_2$	$Q\ B_2$
	5	$Q\ A_2$	$Q\ B_2$	$P\ A_1$	$P\ B_1$	$R\ A_3$	$R\ B_3$
	6	$R\ A_3$	$R\ B_3$	$Q\ A_2$	$Q\ B_2$	$P\ A_1$	$P\ B_1$

crop-treatment sequences of such an experiment over the two preliminary years and six experimental years. Each of the sequences will be replicated three times, there being 9 blocks in all of two plots each. The yields for the six experimental years for crop P will consist of three replicates of Table 3, where a_1, b_1, a_2, b_2 etc. represent the yields of

TABLE 3. YIELDS OF ONE REPLICATE OF CROP P FROM THE ROTATION EXPERIMENT OF TABLE 2

	Series					
	I		II		III	
Year	1	2	3	4	5	6
1	a_1	b_1				
2			a_2	b_2		
3					a_3	b_3
4	a_4	b_4				
5			a_5	b_5		
6					a_6	b_6

crop P in years 1, 2 etc. and in treatment cycles A and B (suffices being used to denote years). The results for the other two crops will be similar.

The analysis in this case is simple. The yields of crop P in series I will constitute a $2 \times 2 \times 3$ table of treatments $\times$ years (1 and 4) $\times$ replicates (blocks). The degrees of freedom of the analysis of variance of series I can therefore be partitioned as follows:

Treatments (T)	1
Years (Y)	1
Blocks (B)	2
$T \times Y$	1
$T \times B$	2
$Y \times B$	2
$T \times Y \times B$	2
	11

Since the treatments are repeated on the same plots the components T, B, and $T \times B$ which are derived from plot totals over the two years constitute the plot total part of the analysis, the remaining terms the plot $\times$ years part.

The contrasts between the totals of series I, II and III are estimates of year differences (though subject to greater errors than the other components of years). The three single degrees of freedom for treatments from the three partial analyses combine into one for treatments and two for treatments × series (i.e. treatments × years). The combined analysis of the three series for one crop is therefore as shown in Table 4.

TABLE 4. ANALYSIS OF VARIANCE

		D.F.	M.S.	Expectation
	Series (Years)	2		
Plot Totals	Treatments (T)	1		
	Treatments × Series = $T \times Y$	2		
	Blocks (B)	6		
	$T \times B$ (plot error)	6	E_p	$2\sigma_p^2 + \sigma_w^2$
		15		
Plots × Years	Years (Y)	3		
	$T \times Y$	3		
	$Y \times B$	6		
	$T \times Y \times B$ (plot × year error)	6	E_w	σ_w^2
		18		
		35		

The $T \times B$ component with 6 d.f. provides an estimate of error (which we may term *plot error*) for comparisons based on plot totals over all years, and the $T \times Y \times B$ component, also with 6 d.f., provides an estimate (which we may term *plot × year error*) for comparisons not involving differences between plots. The estimate of the error variance of the difference between A and B averaged over all years, for example, will be $\frac{1}{9}E_p$, and that of the change in this difference between the first three years and the second three years will be $\frac{4}{9} E_w$.

If the errors of the plot yields are regarded as made up of two components, one independent from year to year with variance σ_w^2 and the other constant over all years with variance σ_p^2, the expectation of E_w is σ_w^2 and of E_p is $2\,\sigma_p^2 + \sigma_w^2$. (The coefficient of σ_p^2 is given by the number of yearly yields, two in this case, entering into each plot total.)

The sum of the sums of squares for the two estimates of error will be equal to the sum of the sums of squares for error in the analyses of the results of each year separately. There will be 2 d.f. for error in each year giving 12 d.f. in all, and the expectation of each mean square will be $\sigma_p^2 + \sigma_w^2$, agreeing with the combined expectation of $T \times Y$ and $T \times Y \times B$.

If results are available for a larger multiple of three years the first part of the analysis will be unaltered except for change of divisors, but the degrees of freedom in the second part will be increased. With 12 years, for example, all these degrees of freedom will be multiplied by 3, and the expectation of E_p will be $4\,\sigma_p^2 + \sigma_w^2$. The analysis is also easily extended to an experiment containing more than two treatment sequences. With t treatment sequences a factor $t - 1$ will be introduced into the degrees of freedom of all components involving T. The alterations for change of rotation length are equally simple.

In experiments extending over a long period of time it is sometimes advisable, when trends are being considered, to subdivide the plots $\times$ years part of the analysis further, into say linear trend and remainder, so as to take account of possible differential trends of plots within blocks. This point is discussed by Cochran (1939) and Patterson (1953) and need not detain us here.

It will be noted that the interaction of years with blocks, $Y \times B$, has not been included in the estimate of the plot $\times$ year component of error. This is as it should be, since the different blocks may well exhibit year to year differences: such differences will not enter into the treatment comparisons since these are all compounded of differences within blocks in the same years. It is the elimination of this component which complicates the analysis of the rice-pasture experiment which is the subject of our investigation.

A rotation in which the same crop occurs more than once

If the same crop occurs more than once in a rotation comparisons can only be made with precision between the yields of the crop at the different stages of the rotation if the plots carrying them are arranged together in blocks. This requires that some at least of the different phases of the rotation must be grouped together in the same blocks. Whether all phases have to be so grouped depends on the crop rotation. In the four course rotation potatoes, wheat, potatoes, barley, for example, phases 1 and 3 must be grouped together, as must phases 2 and 4. In the rotation potatoes, potatoes, barley, wheat all four phases must be grouped together.

Whether comparisons between the same crop at different stages of a rotation are required will depend on the nature of the experiment. Such comparisons are usually of some interest, but may not be judged of sufficient value to compensate for the loss of accuracy resulting from the larger blocks and the inconvenience and other troubles arising from the growing of more than one crop in the same block. We shall not discuss the possibilities here, as the problems are essentially those of design.

When the different phases of a rotation are grouped together new problems of analysis arise. We will first consider the simple example of a three-course rotation with two treatment cycles A and B as before, but with one crop repeated in two consecutive years e.g. with the rotation of crops P.P.R. instead of the rotation P.Q.R. of Table 1. Each replicate of the three phases of the rotation will be taken to be arranged together in a block instead of in three separate blocks. The yields of one replicate of crop P will now be as shown in Table 5.

TABLE 5. YIELDS OF THE DUPLICATED CROP IN ONE REPLICATE OF A THREE COURSE ROTATION WITH TWO CROPS THE SAME

	Sequence					
Year	1	2	3	4	5	6
1	a_1	b_1			a'_1	b'_1
2	a'_2	b'_2	a_2	b_2		
3			a'_3	b'_3	a_3	b_3
4	a_4	b_4			a'_4	b'_4
5	a'_5	b'_5	a_5	b_5		
6			a'_6	b'_6	a_6	b_6

Here a'_1 , b'_1 , etc. represent the yields of the second P crop under treatment cycles A and B. The contrast between $\frac{1}{2}(\bar{a} + \bar{b})$ and $\frac{1}{2}(\bar{a}' + \bar{b}')$ is a *crop-phase* contrast. $\frac{1}{2}(\bar{a} - \bar{b} - \bar{a}' + \bar{b}')$ will then represent the treatment $\times$ crop-phase interaction.

The comparisons of chief interest will be between the means over all years of a, a', b and b'. The difference between $\frac{1}{2}(\bar{a} + \bar{a}')$ and $\frac{1}{2}(\bar{b} + \bar{b}')$ is derived from plot (i.e. sequence $\times$ replicate) totals over all years. The differences between $\bar{a}$ and $\bar{a}'$, and between $\bar{b}$ and $\bar{b}'$ do not involve plot differences.

If the same crop were grown in all years the interaction component of the (6×3) plot total $\times$ replicate table would represent plot error. In the present case, however, the plot totals over the six years will be affected by year differences. Sequences 1 and 2 contain years 1, 2, 4 and 5, sequences 3 and 4 years 2, 3, 5 and 6, etc. Hence the interaction component of the plot total $\times$ replicate table will contain components of block × year interactions, which as pointed out above, require to be eliminated from error.

A partial analysis is easily effected. If plots 1 and 2, and the corresponding plots in the other replicates, are taken, we obtain a 2 × 4 × 3 table of treatments × years × replicates, phases being completely confounded with years. The interaction of the (2 × 3) plot total × replicates table gives 2 d.f. for plot error, and the three factor interaction gives 6 d.f. for plot × year error. Similar analyses can be made for plots 3 and 4 and plots 5 and 6. Combining the results we obtain an estimate of plot error with 6 d.f. and an estimate of plot × year error with 18 d.f.

In experiments with a fair number of treatments this partial analysis may be judged sufficient to give adequate estimates of error, but it is of interest to consider how the remaining degrees of freedom for error can be recovered. In experiments containing rotations of varying length or with variations in rotations of the same length, without other treatments, as in the rice-pasture experiment under consideration, recovery of these degrees of freedom is essential.

The number of degrees of freedom for error from a single year's results will be 6. Hence there will be 36 d.f. in all for error. The full table of plot totals will consist of a 6 × 3 table of sequences × blocks. This indicates that the total number of degrees of freedom for plot error is 5 × 2 = 10. Hence there will be 26 d.f. for plot × year error. We therefore require an additional 4 d.f. for plot error and an additional 8 d.f. for plot × year error.

If c_1 is written for $a_1 + b_1$, c_1' for $a_1' + b_1'$ etc. the yields of one replicate can be written as shown in Table 6.

TABLE 6. COMBINED YIELDS OF ONE REPLICATE

	Sequences			
Year	1 and 2	3 and 4	5 and 6	Total
1	c_1	·	c_1'	Y_1
2	c_2'	c_2	·	Y_2
3	·	c_3'	c_3	Y_3
4	c_4	·	c_4'	Y_4
5	c_5'	c_5	·	Y_5
6	·	c_6'	c_6	Y_6
Total	$P_{1,2}$	$P_{3,4}$	$P_{5,6}$	R

Only the differences within columns of the c's and c''s, which have appeared as year differences, have so far been taken into account. We shall now consider the full analysis of this table.

There are 11 d.f. in all which can be partitioned into

Columns (plot pair totals)	2
Years	5
Columns × Years	4

Columns and years are not, however, orthogonal. To overcome this we may replace the column totals $P_{1,2}$ etc. by the quantities Q_1, Q_2, Q_3, given by

$$Q_1 = 2\{c_1 + c_2' + \tfrac{1}{2}(c_3 + c_3') + c_4 + c_5' + \tfrac{1}{2}(c_6 + c_6')\}$$
$$= 2P_{1,2} - Y_{(1)} + R$$

etc., where $Y_{(1)}$ represents the total $Y_1 + Y_2 + Y_4 + Y_5$ of the year × replicate totals for the years in which sequence 1 (or 2) carries the crop. All years are equally represented in these quantities, which are therefore orthogonal with years. They also represent differences between multiples of plot totals. Consequently if the quantities are calculated for all three replicates the interaction of the resultant 3 × 3 table of Q's will give 4 d.f. for plot error.

The divisor for the squares of the Q's, and the expectation of the error in terms of σ_p^2 and σ_w^2, remain to be determined.

For the divisor we have

$$Q_1 - Q_2 = 2\{c_1 - \tfrac{1}{2}(c_1 + c_1') + c_2' - c_2 + \tfrac{1}{2}(c_3 + c_3') - c_3' + \cdots\}$$
$$= (c_1 - c_1') + 2(c_2' - c_2) + (c_3 - c_3') + \cdots$$

Since each c is a total of two plots the divisor of $(Q_1 - Q_2)^2$ is $2\{1^2 \times 8 + 2^2 \times 4\} = 48$. The divisor of each Q^2 is therefore 24.

To evaluate the expectation in terms of σ_p^2 and σ_w^2 we replace all yields of plot 1 by p_1, etc. We then have

$$Q_1 - Q_2 = 6p_1 + 6p_2 - 6p_3 - 6p_4$$

Hence the expectation of the σ_p^2 component of $(Q_1 - Q_2)^2$ is $4 \times 6^2 \times \sigma_p^2 = 144\sigma_p^2$. The expectation of the corresponding error mean square is therefore $3\sigma_p^2 + \sigma_w^2$. If years and plots had been orthogonal the expectation would have been $4\sigma_p^2 + \sigma_w^2$.

Alternatively the sum of the coefficients of the p^2 terms in $Q_1^2 + Q_2^2 + Q_3^2 - \tfrac{1}{3}(Q_1 + Q_2 + Q_3)^2$ can be calculated.

If these 4 d.f. for plot error and the 6 d.f. already obtained are combined by adding the sums of squares the expectation of the resultant mean square (10 d.f.) will be

$$\frac{4}{10}(3\sigma_p^2 + \sigma_w^2) + \frac{6}{10}(4\sigma_p^2 + \sigma_w^2) = 3.6\sigma_p^2 + \sigma_w^2$$

This method of combination is only fully efficient if σ_p^2 is small relative to σ_w^2. Otherwise greater weight should theoretically be given to the mean square with larger expectation. Even in the extreme case when σ_p^2 is very large relative to σ_w^2, however, the loss of information with the above weighting is quite trivial. Nevertheless, as will be seen later, the combined estimate is by no means equivalent to an estimate with 10 d.f. and expectation $4\sigma_p^2 + \sigma_w^2$.

The analysis of the three replicates of the c's can now be completed from the table of the totals of the c's over all replicates. It will take the form shown in Table 7, where the phase contrast $\bar{c} - \bar{c}'$ is represented by P.

TABLE 7. ANALYSIS OF VARIANCE OF TABLE 5

	d.f.	
1. Years (ignoring treatments)	5	From the block × year totals table
2. Blocks (replicates)	2	
3. $Y \times B$ (ignoring treatments)	10	
4. Q's: totals over blocks ($P \times Y$)	2	As above
5. Q's: interaction with blocks (plot error)	4	
6. Remainder: totals over blocks (P, 1; $P \times Y$, 3)	4	By differences
7. Remainder: interaction with blocks (plot × year error)	8	
	35	

The sums of squares for the last two items are obtained by differences. The total sum of squares for the table of the totals of the c's over all replicates equals the total of the sums of squares for items 1, 4 and 6. The total of the sums of squares for all seven items is equal to the sum of squares for the 36 c's. Item 7 gives the remaining 8 d.f. for plot × year error.

It will be seen that the key to this analysis is given by the expressions Q which make the plot pair (column) totals orthogonal with years. In this example these expressions can be written down by inspection, but where this cannot be done they can be obtained by fitting constants for years (rows) and columns by the method of least squares. In the next section the method will be illustrated by application to the rice-pasture experiment.

The full analysis can now be completed. If required T, P, $T \times P$ and their interactions with years can be separately exhibited. It then takes the form shown in Table 8.

TABLE 8. FULL ANALYSIS OF VARIANCE

	D.F.	Expectation		D.F.	Expectation
T	1	$4\ \sigma_p^2 + \sigma_w^2$	Years (ignoring treatments)	5	
P	1	σ_w^2	Blocks	2	
$T \times P$	1	σ_w^2	Y $\times B$ (ignoring treatments)	10	
$T \times Y$	5	$0.4\ \sigma_p^2 + \sigma_w^2$	Plot error	10	$3.6\ \sigma_p^2 + \sigma_w^2$
$P \times Y$	5	$1.2\ \sigma_p^2 + \sigma_w^2$	Plot $\times$ years error	26	σ_w^2
$T \times P \times Y$	5	$1.2\ \sigma_p^2 + \sigma_w^2$		71	

T, P, $T \times P$, $T \times Y$ and $T \times P \times Y$ are computed in the ordinary manner. $P \times Y$ is obtained by subtraction of P from items 4 and 6 of Table 6. The errors of the various comparisons and the expectations of mean squares such as $T \times Y$ can be evaluated in terms of σ_p^2 and σ_w^2 by replacing yields by plot constants in the manner already indicated. The mean expectation of the 18 d.f. for treatments and treatments $\times$ years checks to $\sigma_p^2 + \sigma_w^2$, as it should.

It should be noted the plot error mean square no longer gives the error of differences based entirely on plot totals. Thus the error variance of the treatment difference $\frac{1}{2}(\bar{a} + \bar{a}') - \frac{1}{2}(\bar{b} + \bar{b}')$ is $\frac{2}{9}(4\sigma_p^2 + \sigma_w^2)$. The estimate of this will be $\frac{2}{9}(\frac{10}{9}E_p - \frac{1}{9}E_w)$. The accuracy of this estimate of error, relative to the direct estimate based on the 6 d.f. of the partial analysis is about equivalent to an extra 2 d.f. when σ_p^2 is small relative to σ_w^2, but equivalent to nearly an extra 4 d.f. when σ_p^2 is large relative to σ_w^2.

The analysis of the other crop of the rotation will be similar to the analysis of Table 4. There will now only be 2 d.f. for blocks, the remaining 4 d.f. representing year $\times$ block interactions. There will still only be 6 d.f. for plot $\times$ year error, since the other 4 d.f. will be completely confounded with year $\times$ block interactions.

Orthogonal partition of the degrees of freedom in the rice-pasture experiment.

In the rice-pasture experiment orthogonal Q functions for the 11 crop sequences cannot be written down by inspection. Fitting of constants must therefore be resorted to. The required functions can be obtained from the equations appropriate to a single replicate.

The yields and year and plot totals of one replicate are shown in Table 1. Let the year constants be $y_1, y_2 \cdots y_{12}$, and the plot constants $p_1, p_2, \cdots p_{11}$. The normal equations are then:

$$4y_1 + p_1 + p_4 + p_7 + p_8 = Y_1$$
$$4y_2 + p_2 + p_4 + p_5 + p_9 = Y_2$$
$$\cdots\cdots\cdots\cdots$$
$$4p_1 + y_1 + y_4 + y_7 + y_{10} = P_1$$
$$\cdots\cdots\cdots\cdots$$
$$6p_4 + y_1 + y_2 + y_5 + y_6 + y_9 + y_{10} = P_4$$
$$\cdots\cdots\cdots\cdots$$
$$3p_8 + y_1 + y_5 + y_9 = P_8$$
$$\cdots\cdots\cdots\cdots$$

Denote the sum of the Y's for the years in which plot 1 carries rice by $Y_{(1)}$, so that $Y_{(1)} = Y_1 + Y_4 + Y_7 + Y_{10}$, etc. Further put $p_1 + p_2 + p_3 = p_{1-3}$, $p_4 + p_6 = p_{4,6}$, $p_5 + p_7 = p_{5,7}$, $p_8 + p_9 + p_{10} + p_{11} = p_{8-11}$, with similar totals for the P's.

Eliminating the y's in the p equation by means of the y equations we then have:

$$12p_1 - 2p_{4,6} - 2p_{5,7} - p_{8\text{-}11} = 4P_1 - Y_{(1)} \quad (1)$$
$$12p_2 - 2p_{4,6} - 2p_{5,7} - p_{8\text{-}11} = 4P_2 - Y_{(2)} \quad (2)$$
$$12p_3 - 2p_{4,6} - 2p_{5,7} - p_{8\text{-}11} = 4P_3 - Y_{(3)} \quad (3)$$
$$18p_4 - 2p_{1\text{-}3} - 3p_{5,7} - 3p_8 - 3p_9 = 4P_4 - Y_{(4)} \quad (4)$$
$$18p_5 - 2p_{1\text{-}3} - 3p_{4,6} - 3p_9 - 3p_{10} = 4P_5 - Y_{(5)} \quad (5)$$
$$18p_6 - 2p_{1\text{-}3} - 3p_{5,7} - 3p_{10} - 3p_{11} = 4P_6 - Y_{(6)} \quad (6)$$
$$18p_7 - 2p_{1\text{-}3} - 3p_{4,6} - 3p_8 - 3p_{11} = 4P_7 - Y_{(7)} \quad (7)$$
$$9p_8 - p_{1\text{-}3} - 3p_4 - 3p_7 = 4P_8 - Y_{(8)} \quad (8)$$
$$9p_9 - p_{1\text{-}3} - 3p_4 - 3p_5 = 4P_9 - Y_{(9)} \quad (9)$$
$$9p_{10} - p_{1\text{-}3} - 3p_5 - 3p_4 = 4P_{10} - Y_{(10)} \quad (10)$$
$$9p_{11} - p_{1\text{-}3} - 3p_6 - 3p_7 = 4P_{11} - Y_{(11)} \quad (11)$$

We may now take the following sums of these equations:

(1) + (2) + (3)

$$12p_{1\text{-}3} - 6p_{4,6} - 6p_{5\text{-}7} - 3p_{8\text{-}11} = 4P_{1\text{-}3} - R$$

3(4) + 3(6) + (5) + (7)

$$36p_{4,6} - 16p_{1\text{-}3} - 12p_{8\text{-}11} = 12P_{4,6} + 4P_{5,7} - 4R$$

$3(5) + 3(7) + (4) + (6)$

$$36p_{5,7} - 16p_{1-3} - 12p_{8-11} = 12P_{5,7} + 4P_{4,6} - 4R$$

$(8) + (9) + (10) + (11)$

$$9p_{8-11} - 4p_{1-3} - 3p_{4,6} - 3p_{5,7} = 4P_{8-11} - R$$

It is easily verified that these equations are satisfied by

$$\left.\begin{aligned} p_{1-3} &= \tfrac{1}{4}P_{1-3} \\ p_{4,6} &= \tfrac{1}{6}P_{4,6} \\ p_{5,7} &= \tfrac{1}{6}P_{5,7} \\ p_{8-11} &= \tfrac{1}{3}P_{8-11} \end{aligned}\right\} \qquad (12)$$

This is a consequence of the fact that the sums of plots 1–3, 4 and 6, 5 and 7, and 8–11 are all orthogonal with years.

The values of the separate p's are then obtained by substitution in equations (1)–(11), and elimination of $p_8 - p_{11}$ from equations (4)–(7) by means of equations (8)–(11). This gives

$$\left.\begin{aligned} 12p_1 &= 4P_1 - Y_{(1)} - \tfrac{1}{3}P_{1-3} + \tfrac{1}{3}R \\ &\cdots\cdots\cdots\cdots\cdots \end{aligned}\right\} \qquad (13)$$

$$\left.\begin{aligned} 24p_4 &= 5P_4 - P_6 + P_8 + P_9 - P_{10} - P_{11} - Y_{(4)} + Y_{(6)} \\ &\cdots\cdots\cdots\cdots\cdots \end{aligned}\right\} \qquad (14)$$

$$\left.\begin{aligned} 36p_8 = 17P_8 - P_{10} + 3P_4 + 3P_7 - 5Y_{(8)} + Y_{(10)} + P_{1-3} \\ - \tfrac{1}{2}P_{4,6} - \tfrac{1}{2}P_{5,7} \\ \cdots\cdots\cdots\cdots\cdots \end{aligned}\right\} \qquad (15)$$

The total sum of squares accounted for by fitting the constants might be calculated from the expression $S(yY) + S(pP)$, the values of the year constants y being determined by substitution of the p values in the original normal equations, or alternatively from the expression $\frac{1}{4}S(Y^2) + S(pP')$, when $4P_1' = 4P_1 - Y_{(1)}$ etc. It is more satisfactory, however, to construct a set of orthogonal Q functions which partitions the sum of squares directly.

Such a set of functions can be constructed by taking appropriate groups of constants in turn in any desired order, and obtaining expressions for these in terms of the Y's and P's after eliminating all groups of constants already fitted. If certain conditions of symmetry are satisfied the sum of squares attributable to a group of constants will be calculable from the sum of the squares of the deviations of the values of the corresponding Q functions in the group.

As a first step we may replace the p's by $\bar{p}_{1-3}$, $\bar{p}_{4,6}$, $\bar{p}_{5,7}$ and $\bar{p}_{8-11}$, δp_1 , δp_2 , $\cdots$ δp_{11} , where $\bar{p}_{1-3} = \frac{1}{3}p_{1-3}$, $\bar{p}_{4,6} = \frac{1}{2}p_{4,6}$, etc. and $\delta p_1 = p_1 - \bar{p}_{1-3}$, $\delta p_4 = p_4 - \bar{p}_{4,6}$, etc. We may then take the following groups of constants in the order shown:

1. y_1 , y_2 , $\cdots$ y_{12}
2. $\bar{p}_{1-3}$, $\bar{p}_{4,6}$, $\bar{p}_{5,7}$, $\bar{p}_{8-11}$
3. δp_1 , δp_2 , δp_3
4. δp_8 , δp_9 , δp_{10} , δp_{11}
5. δp_4 , δp_6 and δp_5 , δp_7

The ordinary expressions of the analysis of variance give the sums of squares accounted for by groups 1 and 2, which are orthogonal.

The values of δp_1 , δp_2 , δp_3 before fitting δp_4 to δp_{11} are given by replacing p_4 by $\bar{p}_{4,6}$ etc. in equations (1)–(3). It will be seen that the equations (13) already obtained are unaltered, and the corresponding sum of squares will be given by

$$\frac{1}{\lambda}\,\text{dev}^2\{Q_1 , Q_2 , Q_3\}$$

where $Q_1 = 4P_1 - Y_{(1)}$ etc., dev^2 indicates the sum of the squares of the deviations, and λ is a divisor to be determined.

We have, by ordinary least squares procedure,

$$V(\delta t_1 - \delta t_2) = (c_{11} - 2c_{12} + c_{22})\sigma^2$$

where c_{11} and c_{12} are the values of δt_1 and δt_2 when $P_1 = +\frac{2}{3}$, $P_2 = P_3 = -\frac{1}{3}$, $P_{4,6} = P_{5,7} = P_{8-11} = 0$ and all $Y = 0$, and c_{21} $(= c_{12})$ and c_{22} are the values of δt_1 and δt_2 when $P_2 = \frac{2}{3}$, $P_1 = P_3 = -\frac{1}{3}$, $P_{4,6} = P_{5,7} = P_{8-11} = 0$, and all $Y = 0$. From equations (13)

$$12\ \delta p_1 = 4P_1 - Y_{(1)} - \frac{4}{3}P_{1\text{-}3} + \frac{1}{3}R$$

Thus

$$c_{11} = c_{22} = \frac{1}{12}\cdot\frac{8}{3} = \frac{2}{9}$$

$$c_{12} = -\frac{1}{12}\cdot\frac{4}{3} = -\frac{1}{9}$$

and

$$c_{11} - 2c_{12} + c_{22} = \frac{2}{3}$$

Hence

$$V(\delta t_1 - \delta t_2) = V\frac{1}{12}(Q_1 - Q_2) = \frac{2}{3}\sigma^2$$

Hence

$$\lambda = \frac{1}{2}\cdot 12^2 \cdot \frac{2}{3} = 48.$$

The values for δp_8 to δp_{11} before fitting δp_4 to δp_7 are given by equations (8)–(11), replacing p_4 and p_6 by $\bar{p}_{4,6}$ and p_5 and p_7 by $\bar{p}_{5,7}$. Hence

$$9p_8 = 4P_8 - Y_{(8)} + \tfrac{1}{4}P_{1\text{-}3} + \tfrac{1}{4}P_{4,6} + \tfrac{1}{4}P_{5,7}$$

By putting $P_8 = +\frac{3}{4}$, $P_9 = P_{10} = P_{11} = -\frac{1}{4}$ and all other P and all $Y = 0$ we obtain $c_{88} = +\frac{1}{3}$, $c_{89} = -\frac{1}{9}$, and therefore $c_{88} - 2c_{89} + c_{99} = \frac{8}{9}$. Hence the divisor is 36.

Finally the values for δp_4 to δp_7 after fitting all other constants are given by equations (14). We have

$$12(\delta p_4 - \delta p_6) = 3(P_4 - P_6) + P_8 + P_9 - P_{10} - P_{11} - Y_{(8)} - Y_{(9)} + Y_{(10)} + Y_{(11)}$$

By putting $P_4 = +\frac{1}{2}$, $P_6 = -\frac{1}{2}$ and all other P and all $Y = 0$ we obtain

$$c_{44} - c_{46} = \tfrac{1}{4}$$

and therefore $c_{44} - 2c_{46} + c_{66} = \frac{1}{2}$. Hence the divisor of 144 $(\delta p_4 - \delta p_6)^2$ is 72.

We may therefore summarize the analysis of a single replicate as follows:

In addition to the year totals, $Y_1 - Y_{12}$, and the totals of the four groups of plots orthogonal with years P_{1-3}, $P_{4,6}$, $P_{5,7}$ P_{8-11}, calculate the following 11 quantities $Q_1 - Q_{11}$.

For plots 1–3: $Q_1 = 4P_1 - Y_{(1)}$, etc. Total: $4P_{1-3} - R$

" " 4–7: $Q_4 = 3P_4 + P_8 + P_9 - Y_{(8)} - Y_{(9)}$ etc.

Total: $3P_{4,6} + 3P_{5,7} + 2P_{8-11} - 2R$

" " 8–11: $Q_8 = 4P_8 - Y_{(8)}$, etc. Total: $4P_{8-11} - R$

The expressions for the sums of squares are shown in Table 9.

The reason why the groups were taken in the chosen order will now be clear. If $\delta p_8 - \delta p_{11}$ had been taken after $\delta p_4 - \delta p_7$ equations (15) would have been used for determining c_{88}, c_{89}, etc. In this case $c_{8,9}$ would not be equal to $c_{8,10}$, etc., i.e. not all the covariances between $\delta p_8 - \delta p_{11}$ would be equal. Consequently the expression for the sum

TABLE 9. ANALYSIS OF VARIANCE: EXPRESSIONS FOR SUMS OF SQUARES

		D.F.	S.S.
Years		11	$\frac{1}{4}\,\text{dev}^2(Y_1, Y_2, \cdots Y_{12})$
Plot Totals	Orthogonal groups	3	$\frac{1}{12}\,\text{dev}^2(P_{1\text{-}3}, P_{4,6}, P_{5,7}, P_{8\text{-}11})$
	Within groups 1–3	2	$\frac{1}{48}\,\text{dev}^2(Q_1, Q_2, Q_3)$
	" groups 4, 6 and 5, 7	2	$\frac{1}{72}\,\{(Q_4 - Q_6)^2 + (Q_5 - Q_7)^2\}$
	" group 8–11	3	$\frac{1}{36}\,\text{dev}^2(Q_8, Q_9, Q_{10}, Q_{11})$
	Total	10	
Remainder (plots × years)		26	
		47	

of squares for $\delta p_8 - \delta p_{11}$ would be more complicated, without any compensating simplification in the sum of squares for $\delta p_4 - \delta p_7$.

The divisors in the above sums of squares may be checked by setting out the coefficients of the differences of pairs of constants and summing their squares. Thus $9(\delta p_8 - \delta p_9)$ has the coefficients shown in Table 10, with sum of squares 72.

TABLE 10. COEFFICIENTS OF $9(\delta p_8 - \delta p_9)$

Year	1	2	3	4	5	6	7	8	9	10	11	Total
1	−1	·	·	−1	·	·	−1	+3	·	·	·	0
2	·	+1	·	+1	+1	·	·	·	−3	·	·	0
5	·	−1	·	−1	·	·	−1	+3	·	·	·	0
6	·	·	+1	+1	+1	·	·	·	−3	·	·	0
9	·	·	−1	−1	·	·	−1	+3	·	·	·	0
10	+1	·	·	+1	+1	·	·	·	−3	·	·	0
Total	0	0	0	0	+3	0	−3	+9	−9	0	0	0

Similarly the orthogonality of the different components can be checked by verifying that the sum of the products of the corresponding coefficients of the differences of each pair of constants is zero.

An alternative numerical check of the divisors and of orthogonality can be made by assuming values of the year and plot constants, building up the plot yields from them, and carrying out the analysis. If the expressions are correct the residual sum of squares (26 d.f.) should be zero. This provides a useful overall check and a test of the numerical procedure, though it should be remembered that a single numerical test does not necessarily detect all errors. Taking $y_1 = 1, y_2 = 2, \cdots y_{12} = 12$, and $p_1 = 1, p_2 = 2, p_{11} = 11$, the yields of plot 1 are 2, 5, 8, 11, those of plot 2 are 4, 7, 10, 13, etc. and the analysis shown in Table 11 is obtained.

TABLE 11. NUMERICAL CHECK

	D.F.	S.S.
Years	11	659.75
Orthogonal groups	3	344.25
Q_1, Q_2, Q_3	2	6
$Q_4 - Q_6, Q_5 - Q_7$	2	16
Q_8, Q_9, Q_{10}, Q_{11}	3	7.25
	21	1033.25

The total agrees with the total sum of squares of deviations of all yields (47 d.f.)

Scheme of analysis of the rice-pasture experiment

We may now complete the scheme of analysis for the rice-pasture experiment. Table 12 shows the general scheme, which has the same pattern as Table 8.

TABLE 12. ANALYSIS OF VARIANCE OF THE RICE-PASTURE EXPERIMENT

		D.F.
Plot Totals	Crop sequences	10
	Blocks	2
	Plot error	20
	Total	32
	Years	11
Plots × Years	Crop sequences × years	26
	Blocks × years	22
	Plot × year error	52
	Total	100
		143

Tables corresponding to Table 1 must be prepared for each replicate, and for the total of all three replicates. The sums of squares for blocks, years and blocks $\times$ years are computed in the ordinary manner from the marginal totals for years. For crop sequences and plot error the quantities P_{1-3} , $P_{4,6}$, $P_{5,7}$, P_{8-11} and $Q_1 - Q_{11}$ must be calculated for each replicate and for the total of all replicates. These are set out in four tables, namely that for the P totals and those for Q_1 to Q_3 , Q_4 to Q_7 , and Q_8 to Q_{11} . Columns of the differences $Q_4 - Q_6$ and $Q_5 - Q_7$ are also required. The marginal totals for all replicates (with an additional factor 3 in the divisors) give the components of the sum of squares for crop sequences (10 d.f.) while the interaction components (which for $Q_4 - Q_6$ and $Q_5 - Q_7$ are merely the sums of the squares of the deviations) give the components of the sum of squares for plot error (20 d.f.).

The sum of squares for crop sequences $\times$ years (26 d.f.) is then obtained by subtracting the sums of squares for years and for crop sequences from the total sum of squares (47 d.f.) for the crop sequence $\times$ years table (totals over all replicates). Likewise the sum of squares for plot $\times$ year error (52 d.f.) is obtained by subtracting all other items from the sum of squares (143 d.f.) for the whole experiment.

It remains to determine the expectations of the various components of the plot error sums of squares in terms of σ_p^2 and σ_w^2 . The procedure already explained of evaluating the sums of squares of the coefficients of the plot constants is followed. The group 8–11 component can be broken down into 3 orthogonal components. We may take the orthogonal components $9(\delta p_8 - \delta p_9)$, $9(\delta p_{10} - \delta p_{11})$ and $9(\delta p_8 + \delta p_9 -$

TABLE 13. EXPECTATIONS OF PLOT-ERROR MEAN SQUARES

		D.F.	Expectation
Orthogonal groups	$P_{4,6} - P_{5,7}$	2	$6\,\sigma_p^2 + \sigma_w^2$
	$P_{1,3} - P_{8,11}$	2	$\frac{7}{2}\,\sigma_p^2 + \sigma_w^2$
	$P_{4,6} + P_{5,7} - P_{1,3} - P_{8,11}$	2	$\frac{19}{4}\,\sigma_p^2 + \sigma_w^2$
Within group 1–3		4	$3\,\sigma_p^2 + \sigma_w^2$
” groups 4, 6 and 5,7		4	$4\,\sigma_p^2 + \sigma_w^2$
” group 8–11	$\delta p_8 - \delta p_9$ and $\delta p_{10} - \delta p_{11}$	4	$\frac{5}{2}\,\sigma_p^2 + \sigma_w^2$
	$\delta p_8 + \delta p_9 - \delta p_{10} - \delta p_{11}$	2	$\frac{11}{4}\,\sigma_p^2 + \sigma_w^2$
Total		20	$3.6\,\sigma_p^2 + \sigma_w^2$

$\delta p_{10} - \delta p_{11}$). The σ_p^2 component of $81(\delta p_8 - \delta p_9)^2$, for example, obtained from the total line of Table 8, is $(3^2 + 3^2 + 9^2 + 9^2)\,\sigma_p^2 = 180\,\sigma_p^2$. Since the divisor is 72 the expectation of the corresponding sum of squares (1 d.f.) is $\frac{5}{2}\sigma_p^2 + \sigma_w^2$. A similar procedure can be followed for the orthogonal groups component. The complete set of expectations is shown in Table 13.

The expectation for the whole 72 d.f. for plot and plot × year error is therefore

	D.F.	M.S.	Expectation
Plot error	20	E_p	$3.6\,\sigma_p^2 + \sigma_w^2$
Plot × year error	52	E_w	σ_w^2
Total	72		$\sigma_p^2 + \sigma_w^2$

This is as it should be, since the 72 d.f. is made up of 6 d.f. for error for each year separately, each with expectation $\sigma_p^2 + \sigma_w^2$. The expectation for plot error could of course be obtained by utilising this fact, but direct evaluation provides a useful check and exhibits the structure of the analysis.

In the above analysis the degrees of freedom for crop sequences and for crop sequences × years have been partitioned in a manner suitable for the separation of plot error and plot × year error. They can also be partitioned so as to isolate the various contrasts which are of experimental interest. The main contrasts are those between rotations A and C and the two phases B and B' of B. The contrasts between A and C and the mean of B and B' are part of the 10 d.f. for crop sequences. The sum of squares (2 d.f.) is given by

$$\frac{1}{36}\,T_{1\text{-}3}^2 + \frac{1}{72}\,T_{4\text{-}7}^2 + \frac{1}{36}\,T_{8\text{-}11}^2 - \frac{1}{144}\,T^2$$

where T_{1-3} represents the total over the three replicates of P_{1-3}, etc., and T the grand total. The corresponding mean square is not, however, directly comparable with the plot error mean square, since its expectation, from the results already given, is $\frac{1}{2}(\frac{7}{2} + \frac{19}{4})\,\sigma_p^2 + \sigma_w^2 = \frac{33}{8}\,\sigma_p^2 + \sigma_w^2$. It must therefore be compared with $\frac{55}{48}\,E_p - \frac{7}{48}\,E_w$.

The contrast between B and B' is wholly within plots and is therefore subject to plot × year error, the error variance being $\frac{1}{12}\,E_w$, and the sum of squares (1 d.f. from crop sequences and years) $\frac{1}{12}\{S(B) - S(B')\}^2$.

The remaining 33 d.f. from crop sequences and crop sequences × years represent the interactions of rotations A and C and the two phases of rotation B with years. The expectation of the corresponding mean square is $\frac{1}{33}(36 - \frac{33}{4})\, \sigma_p^2 + \sigma_w^2 = \frac{37}{44}\, \sigma_p^2 + \sigma_w^2$ and it is therefore compared with $\frac{185}{792} E_p + \frac{607}{792} E_w$.

It should also be noted that blocks × years contains components of plot differences.

Analysis of incomplete data

In long term agricultural experiments an interim analysis of the results is often required. The interim data usually lack the balance that is attained when the experiment has run its full term. In the type of experiment we have just been discussing an exact analysis may then prove excessively laborious since the relevant normal equations will no longer be readily soluble. On the other hand it is unreasonable to tell the experimenter that he must await the completion of the experiment before attempting to draw any conclusions from the results. The statistician must therefore be prepared to determine how far approximate methods of analysis will enable interim reports to be made on long-term experiments in an expeditious manner and without undue labour.

In the case of the rice-pasture experiment there do not appear to be any approximate methods which are very satisfactory. As an example of the problems involved we may consider the analysis of 6 years' data. We may adopt the same notation as previously except that P_1 etc. and T_1 etc. now represent totals over 6 years instead of 12. In the exact analysis on the lines already laid down, the degrees of freedom would partition in the manner shown in Table 14, which corresponds to Table 9.

TABLE 14. ANALYSIS OF VARIANCE OF 6 YEARS' DATA

		D.F.
Plot Totals	Crop sequences	10
	Blocks	2
	Plot error	20
	Years	5
Plots × years	Crop sequences × years	8
	Blocks × years	10
	Plot × year error	16
		71

An estimate of plot error could be built up from the orthogonal groups and the Q functions already obtained. The components derived from the Q functions are now, however, no longer fully orthogonal amongst themselves (though they are still, as is essential, orthogonal with years). This will not seriously affect the estimate of plot error—it has an analogous effect to the inclusion of some degrees of freedom more than once in an estimate of error—but the sum of squares so obtained cannot now be subtracted from the total sum of squares for error to give the plot × year error without serious risk of disturbance due to non-orthogonality.

The simplest procedure, therefore, is to throw the block × year interaction into the estimates of error. The interaction of the table of treatment sequence totals × blocks (appropriate allowance being made for differing numbers of yields in the totals) will then give an estimate of plot error (containing components of block × year interaction) with 20 d.f. The total of the interaction components of the 11 tables of treatment sequences × replicates (the last two of which contribute nothing) will give an estimate of plot × year error (also containing block × year interaction) with 26 d.f. The expectation in terms of σ_p^2 and σ_w^2 of the plot error can be evaluated by the procedure already adopted.

This method of estimation of error will give estimates which are too large if block × year interactions are substantial. This point can be examined by comparing the mean square for block × year interactions, calculated from the table of block × year totals, with its error expectation in terms of σ_p^2 and σ_w^2. If it is considered that the interaction should be eliminated then a full least square analysis, with inversion of the matrix, will have to be undertaken. Construction of orthogonal functions, which will necessarily be complicated, will not be worth while.

REFERENCES

Cochran, W. G. 1939. Long-term agricultural experiments. *J. R. Statist. Soc. Suppl.*, *6*, 104–148.

Crowther, F. and Cochran, W. G. 1942. Rotation experiments with cotton in the Sudan Gezira. *J. Agric. Sci.*, *32*, 390–405.

Patterson, H. D. 1953. The analysis of the results of a rotation experiment on the use of straw and fertilizers. *J. Agric. Sci.*, *43*, 77–88.

Yates, F. 1949. The design of rotation experiments. *Comm. Bur. Soil Sci. T. C.* No. 46

Paper XI

PRINCIPLES GOVERNING THE AMOUNT OF EXPERIMENTATION IN DEVELOPMENTAL WORK

From NATURE
Volume 170, pp. 138–140, 1952

Author's Note

Misunderstanding of the meaning of tests of significance frequently leads experimenters, when attempting to decide whether treatment A or treatment B is to be recommended, to demand that experimentation should be continued until a significant difference is found between the two treatments. This confuses scientific inference with decision making. Paper XI considers the analogous but simpler quantitative problem where a recommendation has to be made on the optimum level of a given treatment on the basis of experimental results. A solution to the problem of alternatives (A or B) was subsequently given by P. M. Grundy, D. H. Rees, and M. J. R. Healy (*J.R. statist. Soc. B*, **18**, 32-35, 1956).

1969 FRANK YATES

PRINCIPLES GOVERNING THE AMOUNT OF EXPERIMENTATION IN DEVELOPMENTAL WORK

By DR. F. YATES, F.R.S.

Rothamsted Experimental Station

GREAT BRITAIN has lagged badly behind the United States in the practical utilization of scientific discoveries. This is reflected in the demand for men of science in the two countries. The recently published Fifth Annual Report of the Advisory Council on Scientific Policy[1], for example, directs attention to the fact that the United States is turning out nearly three times as many men of science in proportion to the labour force as Great Britain is, and is planning further increases, particularly at the higher level. In part this lag in putting scientific discoveries to practical use is attributable to our neglect of the experiments and tests which are necessary in developmental work. These experiments differ markedly from the laboratory experiments required in pure scientific research, since their main function is to establish empirical rules of operation which are applicable under practical conditions. With the present drive for economy there is serious danger that even such facilities as are available for experimental work of this kind will be curtailed or not used to full advantage. It is therefore important to stress that such curtailment will result in much more substantial and immediate losses through failure to determine the best practices. Developmental work must be expanded, not contracted, if we are to survive.

It is perhaps somewhat remarkable that no very precise consideration appears to have been given to the expenditure that is justifiable in developmental work. Instead, experimenters have tended to rely on their intuitive judgment on the accuracy that should be aimed at, and are frequently influenced in this judgment by the demands, often misplaced, for economy. In its essentials the problem is very similar to that of deciding the accuracy required in a sample survey. For this latter problem I put forward the general principle that the accuracy should be such that the sum of the cost of the survey and the expected losses due to errors in the results should be minimized[2]. The same principle can be applied to experimental work. It is indeed here more widely applicable, since it is usually easier in experimental work to assess the losses due to errors of a given magnitude in the results.

In experimental work, however, we have at the outset to consider the type of results that are likely to emerge and the way in which they will be used. In the case of a treatment of the all-or-nothing type we may simply require a decision on whether or not to apply the treatment uniformly to the whole of the material, or to certain categories, already defined, of the material; or we may hope that the experiments themselves will reveal a way of classifying the material into categories for some of which the treatment will be profitable and for others unprofitable. Similarly, in the case of a quantitative treatment, we may require only to determine what uniform level of treatment gives the greatest economic return, or we may hope to divide the material into categories for which different levels of treatment will be most economic, or find some quantitative characteristic of the material which is correlated with the most economic level.

If the discovery of appropriate categories for non-uniform treatment is the aim, more elaborate experiments will often be required. We may also be prepared to sacrifice some of the smaller gains resulting from pushing the simpler type of investigation to its economic limit in the hope of obtaining the much larger gains which will result from the effective use of differential levels of treatment. For the moment, however, we will leave aside these more difficult problems, and consider the simple case in which the whole of the material is to be treated uniformly.

When a decision whether or not to apply a treatment is required application of the above principle is complicated, and we shall not discuss the matter further here. The problem is probably only soluble in terms of fiducial probability, using the sequential approach, and then with difficulty. It is essentially that propounded by Wald in his "Statistical Decision Functions"[3]. When the most economic level has to be determined, on the other hand, the solution is very simple, and reveals a number of features of general interest.

At the outset it will be well to make clear exactly what is meant by the concept of expected loss due to errors in the results. Consider, for example, the question of the use of fertilizer on an agricultural crop. If the whole of the crop is to be treated alike, the most economic level of dressing, which for convenience we will call the *optimum* level, will be the level at which the cost of a further small increment of dressing exactly equals the value of the resultant average increment in response. In the neighbourhood of the optimum the net loss due to departure from the optimum will in general be proportional to the square of the difference of the actual dressing from the optimum. If therefore, instead of the optimum dressing $\hat{x}$ we apply a dressing $\hat{x} \pm \delta x$ there will be a net loss of $\lambda(\delta x)^2$ per acre, where λ is some constant. If $\hat{x}$ is estimated by means of a set of experiments, and the estimate is subject to an error variance $V(\hat{x})$, then the average value of $(\delta x)^2$ over a large number of such sets of experiments will be $V(\hat{x})$. The expected loss due to error in the optimum will therefore be $\lambda V(\hat{x})$. The actual loss may of course be less or greater than the expected loss. An inaccurate set of experiments, for example, may by chance give the correct answer.

If, therefore, the cost C, apart from any 'overhead' or constant component, of an amount q of experimentation is cq, and the error variance of the estimate of the optimum level determined from such experimentation is v/q, the expected loss L will be $\lambda Av/q$, where A is the amount of the material to be treated. We therefore have

$$C + L = cq + \lambda Av/q.$$

This is a minimum when

$$q = \sqrt{(\lambda Av/c)}.$$

In this case

$$C - L = \sqrt{(\lambda Avc)}.$$

In other words, *the most economic amount of experimentation is that in which the cost of the experimentation apart from overheads is equal to the expectation of loss due to errors.*

If an amount A of material has to be treated each year for an indefinite period, and the experiments take a year to carry out, A must be replaced by A/r, where $100r$ is the interest rate. If constancy of conditions cannot be assumed indefinitely and it is decided to repeat the experiment after t years, then A is replaced by

$$\frac{A}{r}\left\{1 - \frac{1}{(1 + r)^t}\right\}.$$

The net loss resulting from undertaking a fraction or multiple f of the most economic amount of experimentation is

$$(f + 1/f - 2)C.$$

If, for example, the cost, apart from overheads, of the most economic amount of experimentation is £4,000, the costs and losses when f has the values of 1, $\frac{1}{2}$ and $\frac{1}{4}$ will be as follows:

f	Cost of experiments (C)	Expected loss (L)	Cost + loss ($C + L$)
1	£4,000	£4,000	£8,000
$\frac{1}{2}$	£2,000	£8,000	£10,000
$\frac{1}{4}$	£1,000	£16,000	£17,000

This table can be looked on as giving the expected additional gains from additional amounts of experimentation in the neighbourhood of the most economic amount. The total gain from the experimentation as a whole cannot, of course, be assessed by this means. In the common case in which the application of the treatment cannot be risked without *some* experimentation, the total gain will be that due to the use of the treatment. This will often be large compared with the marginal gain in the neighbourhood of the most economic amount of experimentation. If, for example, the gain from using the treatment at the correct optimum level is £100,000, and overheads and preliminary fundamental scientific research are costed at £5,000, with other costs as above, the total and net gains will be as follows:

f	Research expenditure (C + £5,000)	Expected total gain (£100,000 − L)	Expected net gain
1	£9,000	£96,000	£87,000
$\frac{1}{2}$	£7,000	£92,000	£85,000
$\frac{1}{4}$	£6,000	£84,000	£78,000

In other words, the return on the first £6,000 of research expenditure is £84,000, on the next £1,000 is £8,000, and on the next £2,000 is £4,000.

If, therefore, experimental resources are scarce there may be a case for not carrying out the full amount of experimentation that can be justified on economic grounds, since the returns on the last few incremental steps are relatively small, and it may therefore be possible to use the immediately available experimental resources more effectively on other problems. However, the experimentation which is required to determine an optimum for uniform treatment is often capable also, if properly planned, of providing a basis for differential treatment. If successful in this respect further large gains may accrue. Consequently, since gains must on the average result from increasing the experimentation to the most economic level for uniform treatment, even if no effective basis for differential treatment is found, there is every justification for building up an experimental organization which is capable of dealing with this volume of experimentation.

There is one further point which is often of considerable practical importance. In many types of experimentation it is somewhat difficult to ensure that experimental conditions fully conform to the conditions under which the treatment will be used in practice. If this is not the case a constant component of error, that is, a bias, will be introduced. An error of this kind will not be eliminated by increasing the amount of experimentation, nor will it be revealed by the variability of the experimental results. Although the gains resulting from increased experimentation will *on the average* be the same whether or not a bias exists, the elimination of the bias may result in much larger gains. Additional experimental effort may therefore in such cases be much better directed to the elimination, reduction or correction of the bias, rather than to the reduction of the relatively small random component of error. To effect this a radically different type of experiment may be required.

As an example of the application of the above principles to a practical situation we may consider the experimental work on the effect of fertilizers on sugar beet, a crop introduced into Britain after the First World War. The early recommendations were based on Continental experience, but after various unco-ordinated experiments had been carried out a co-ordinated series of factorial experiments was started in 1933 and continued to 1949 to test the responses to nitrogen, phosphate and potash. The average number of experiments was twenty-two per year, and the total annual cost, exclusive of overheads, was about £700 per annum at pre-war prices.

As an example of the practical value of these experiments it may be mentioned that the results obtained up to 1939 enabled confident recommendations to be made in 1940 on the manuring of sugar beet under war-time conditions[4]. It was recommended, for example, that a dressing of 0·7 cwt. nitrogen per acre would be optimal, in contrast to the dressing of 0·45 cwt. nitrogen which had previously been recommended[5], and which was, in fact, about optimum at the prices ruling prior to the War. The net gain resulting from the change from 0·45 cwt. nitrogen to 0·7 cwt. nitrogen was about 15*s.* per acre at 1940 prices. If this change were made on the whole of the sugar beet acreage the total net gain would be £300,000 per annum. The actual net gain was probably considerably larger, since under-manuring of sugar beet (and other crops) was common before the War.

The determination of the most economic number of experiments in a situation of this type is complicated, owing to the fact that there is an additional random component of variation in response from year to year due to variation in meteorological conditions, and there is also a possibility of long-term changes due to changes in agricultural conditions. Without going into details, which will be dealt with elsewhere, it may be stated that an analysis of the variation in the nitrogen responses indicates that as far as this fertilizer is concerned about sixty experiments per year, possibly falling to forty per year after the first few years, may be regarded as most economic. In view of the fact that many farmers do not as yet follow at all closely the recommendations emerging from such experiments, the programme can be regarded as about adequate for the simple function of determining optimum uniform dressings. Indeed, from this point of view, the most serious defect of the experiments was the fact that the selection of the sites was not fully random, so that appreciable biases may have been introduced into the results.

The experiments were, however, quite correctly undertaken with the additional and more ambitious aim of investigating possible differences in response on different soil types, and establishing relations between responses and chemical analyses of the soils. In the case of nitrogen, for example, it was confirmed that the responses to nitrogen on fen soils (constituting 10 per cent of the sugar beet acreage) were very small, and would not pay for the cost of nitrogenous fertilizer. This fact is still not fully appreciated by the farmers concerned, 85 per cent of whom, according to a 1945 survey of fertilizer practice, applied nitrogen averaging 0·3 cwt. nitrogen per acre. Omission of this nitrogen would save about £25,000 per annum at present prices, that is, about twenty times the annual cost of the experiments. Considerably larger gains may be expected from the better utilization of the differences in response to phosphate and potash on different soil types and their relations with the chemical analyses of soils. That there is room for substantial economies is obvious when it is realized that the total cost of fertilizers applied to sugar beet is of the order of £3 million per year. For these more ambitious objectives a larger number of experiments would certainly have been economic.

But if the experimental work on sugar beet must be judged to have been scarcely adequate for the purposes for which it was undertaken it is, in fact, the high-light of all experimental work on fertilizers in Great Britain. Sugar beet is the only crop on which any co-ordinated series of modern well-designed factorial experiments has been carried out and the only one for which there has been any attempt to secure a selection of fields for the experimental sites which would be reasonably representative of the whole of the land growing sugar beet. For no other crop is there any adequate series of experiments for which chemical analyses of the soils of the experimental sites have been made, or for which there are any adequate descriptions of these soils. On the more difficult questions of fertilizer practice, such as how to use fertilizers on grassland, and the residual values of phosphate and potash, the amount of co-ordinated experimental work is negligible.

If experimental work in agriculture is to be expanded to a more economic level it will, of course, require a complete reconstruction of the existing machinery for experiments. More work is required not only on fertilizers but also on the many other aspects of crop and livestock production that are capable of simple and exact investigation by empirical experiments. If properly organized, such activities need not interfere with more fundamental scientific research, since technicians rather than research workers are the main requirement. The principal functions of the research workers will be to determine what experiments are worth while, to see that they are properly planned, and to examine the results for relationships which are unknown or only suspected when the experiments are begun.

I have chosen an example from agriculture because this is the field with which I am most familiar, and because it can be demonstrated by the principles set out in this article that the amount of empirical experimentation is here entirely inadequate. The general principles enunciated are of wide application in many other fields, and in particular in many branches of industrial research. The technique of empirical experimentation on highly variable material is relatively new, depending in large part on statistical developments made by British workers during the past thirty years. We have yet to learn how to use this technique to the best advantage, but that it justifies wider and more thorough use there is no question.

REFERENCES

[1] Fifth Annual Report of the Advisory Council on Scientific Policy (1951-1952). (H.M.S.O., London, 1952, Cmd. 8561.)

[2] Yates, F., "Sampling Methods for Censuses and Surveys" (London : Griffin, 1949).

[3] Wald, A., "Statistical Decision Functions" (New York: Wiley, 1950).

[4] Crowther, E.M., and Yates, F., "Fertilizer Policy in Wartime : the Fertilizer Requirements of Arable Crops" (*Emp.J.Exp.Agric.*, **9**, 77–97, 1941).

[5] Ministry of Agriculture and Fisheries, "Arable Crops on the Farm", Bulletin No. 72, H.M.S.O., London, 1937.

PAPER XII

A FRESH LOOK AT THE BASIC PRINCIPLES OF THE DESIGN AND ANALYSIS OF EXPERIMENTS

FROM THE PROCEEDINGS OF THE FIFTH BERKELEY SYMPOSIUM ON MATHEMATICAL STATISTICS AND PROBABILITY
VOLUME IV, pp. 777-790, 1965-66

Author's Note

Paper XII contains a useful exposition of the many subsidiary logical points that arise in the application of tests of significance and the interpretation of experimental results. It gives a more mature judgment than the earlier papers on the precautions that have to be observed, both in design and analysis, to ensure that estimates of error and tests of significance based on the analysis of variance shall be reliable.

As was perhaps to be expected, there has been some protest (though not, so far as I am aware, in print) at my criticism of the fixed and random effects models. In the light of these comments I have given the matter some further thought, and have now added a note, at the end of the paper, which amplifies and extends the arguments given in it. I still believe that the fixed effects model, in the form it is customarily expounded, is a source of confusion rather than enlightenment, and should be dropped. With the rapid accretion of theory in all branches of statistics it is to my mind most important that only what is really useful is retained and taught to students.

1969 FRANK YATES

A FRESH LOOK AT THE BASIC PRINCIPLES OF THE DESIGN AND ANALYSIS OF EXPERIMENTS

F. YATES
ROTHAMSTED EXPERIMENTAL STATION, HARPENDEN

1. Introduction

When Professor Neyman invited me to attend the Fifth Berkeley Symposium, and give a paper on the basic principles of the design and analysis of experiments, I was a little hesitant. I felt certain that all those here must be thoroughly conversant with these basic principles, and that to mull over them again would be of little interest.

This, however, is the first symposium to be held since Sir Ronald Fisher's death, and it does therefore seem apposite that a paper discussing some aspect of his work should be given. If so, what could be better than the design and analysis of experiments, which in its modern form he created?

I do not propose today to give a history of the development of the subject. This I did in a paper presented in 1963 to the Seventh International Biometrics Congress [14]. Instead I want to take a fresh look at the logical principles Fisher laid down, and the action that flows from them; also briefly to consider certain modern trends, and see how far they are really of value.

2. General principles

Fisher, in his first formal exposition of experimental design [4] laid down three basic principles: replication; randomization; local control.

Replication and local control (for example, arrangement in blocks or rows and columns of a square) were not new, but the idea of assigning the treatments at random (subject to the restrictions imposed by the local control) was novel, and proved to be a most fruitful contribution. Its essential function was to provide a sound basis for the assumption (which is always implied in one form or another) that the deviations used for the estimation of error are independent and contain all those components of error to which the treatment effects are subject, and only those components. When a randomized design is used and correctly analyzed disturbances such as those arising from real or imagined fertility gradients in agricultural field trials, and the fact that neighboring plots are likely to be more similar than widely separated plots, can be ignored in the interpretation of the results. The results (yields, and so forth) can in fact be treated as if they were normally and independently distributed about "true" values given by additive constants representing the treatments and block or other local control effects. Valid estimates of the treatment effects and their errors can then be obtained by the classical method of least squares. The analysis of variance (another of Fisher's brilliant contributions) formalizes the arithmetic of this procedure, and permits its extension to more complicated designs, such as split plots, involving a hierarchy of errors.

There is, of course, nothing sacrosanct about the assumptions of normality and additivity, and alternative assumptions can be made if these appear appropriate. What is not sometimes recognized, however, is that the results of one small experiment provide very weak evidence on which to base alternative assumptions. The practical experimenter, or the statistician who works for him, bases his assumptions on long experience of the behavior of the type of material he is handling, and has devices, such as transformations, which enable him to reduce his data to a form which, he is reasonably confident, permits him to apply standard methods of analysis without serious danger of distortion.

This point, I think, Fisher never sufficiently emphasized, particularly as he frequently emphasized the opposing point that each experiment should be permitted to determine its own error, and that no *a priori* information on error should be taken into account.

Nevertheless, Fisher, with his sound practical sense, drew the line between *a priori* and current information where, in the material he was handling, it should be drawn. In other circumstances he would undoubtedly have approved of other methods. When, for example, laboratory determinations of the content of a particular chemical compound are being made on a series of substances, with duplicate determinations on each, that is, 1 d.f. for error for each experiment, determination of error from a control chart, analogous to that used in quality control, is clearly preferable to treating each experiment as an independent entity.

Two points are important here. First, the experimenter does not want to be involved in a haze of indecision on the appropriate methods to apply to the analysis of each particular experiment. Second, when considering the results of an experiment, he should have clearly segregated in his mind the information on the treatment effects provided by the current experiment and that provided by previous experiments with the same or similar treatments.

3. Some points on randomization

There is one point concerning randomization in experiments to which Fisher always appeared to turn a blind eye. As soon as, by some appropriate random process, an experimental layout is determined, the actual layout is known, and can be treated as supplementary information in the subsequent analysis if this appears relevant. Usually, and rightly, the experimenter ignores this information, but there are occasions when it appears wrong to neglect it. Thus, in an agricultural experiment the results may indicate a fertility gradient which is very imperfectly eliminated by the blocks. Should the experimenter not then be permitted to attempt better elimination of this gradient by use of a linear regression on plot position, which can very easily be done by a standard covariance analysis? My own opinion is that when a large and obvious effect of this type is noticed, a statistician would be failing in his duty if he did not do what can be done to eliminate it. But such "doctoring" of the results should be the exception rather than the rule, and when it is resorted to, this should be clearly stated.

Such operations will in general introduce bias into the estimate of error, in the sense that for a fixed set of yields, the average error mean square, for all admissible randomization patterns of dummy treatments, will no longer equal the average treatment mean square. Absence of such bias has been used, by Fisher and others, as one of the justifications for randomization, and as a criterion for the validity of specific randomization procedures. The condition is certainly necessary to ensure full validity of the t and F tests, but can scarcely be regarded as sufficient to ensure a good experimental design. An example is provided by quasi-Latin squares [13]. In this type of design confounding of one set of interactions with the rows of a square, and another set with the columns, enables a 2^6 design, for example, to be arranged in an 8×8 square of plots, row and column differences being eliminated as in a Latin square. The design

2 NP	0 DP	1	3 DN	0 NK	2 DK	1 DNPK	3 PK
0 DNP	2 P	3 D	1 N	2 DNK	0 K	3 NPK	1 DPK
1 D	3 N	2 DNP	0 P	3 DPK	1 NPK	2 K	0 DNK
3	1 DN	0 NP	2 DP	1 PK	3 DNPK	0 DK	2 NK
0 PK	2 DNPK	3 NK	1 DK	2	0 DN	3 DP	1 NP
3 DNK	1 K	0 DPK	2 NPK	1 DNP	3 P	0 N	2 D
1 NK	3 DK	2 PK	0 DNPK	3 NP	1 DP	2 DN	0
2 DPK	0 NPK	1 DNK	3 K	0 D	2 N	1 P	3 DNP

FIGURE 1

4×2^4 factorial design in an 8×8 quasi-Latin square.

is undoubtedly useful, but can with some frequency give an experimental layout in which the contrast between the two pairs of diagonally opposite 4×4 squares represents the main effect of one factor. This I first noticed when exhibiting in 1945 a slide of a 4×2^4 experiment done at Rothamsted in 1939 (figure 1) which had previously been exhibited in 1940 without exciting any comment! This unfortunate contingency is due to the fact that the required confounding is only possible with one rather special basic square. The defect can for the most part be obviated by what I have termed restricted randomization [6]. By excluding the most extreme arrangements, both those of the type mentioned and the complementary type which may be expected on average to be particularly accurate, the unbiased property of the error is preserved.

This throws light on the classical problem of the Knut-Vik or Knight's Move 5×5 Latin square, typified by figure 2(a). This was strongly advocated by

(a)

A	B	C	D	E
D	E	A	B	C
B	C	D	E	A
E	A	B	C	D
C	D	E	A	B

(b)

A	B	C	D	E
E	A	B	C	D
D	E	A	B	C
C	D	E	A	B
B	C	D	E	A

FIGURE 2

Knut-Vik and diagonal 5×5 Latin squares.
(a) Knut-Vik square. (b) Diagonal Latin square.

some as likely to be more accurate than a random Latin square, and Tedin [10] showed by tests on uniformity trials that this was indeed so. The estimate of error, however, is necessarily biased in the opposite direction, so that the results appear *less* accurate than those of a random square. But, of course, a Knut-Vik square *might* be obtained by randomization. If this happens, should the experimenter reject it, and rerandomize? If so, the unbiased property of error will not hold over all remaining squares. This dilemma can in fact be neatly overcome by also excluding the diagonal squares, typified by figure 2(b), which Tedin also investigated and showed to be less accurate than the "random" squares he tested, though because of a small arithmetical error he did not recognize that the loss of accuracy in these squares exactly equals the gain in accuracy in the Knut-Vik squares. This follows immediately from the fact that the four sets of 4 d.f. given by the treatment contrasts of the two Knut-Vik squares and the two diagonal squares are mutually orthogonal, and therefore together comprise the 16 d.f. left after eliminating rows and columns.

From the practical point of view none of this is of great importance, except in such special types of design as quasi-Latin squares. If the experimenter rejects arrangements with obvious systematic features, such as a diagonal pattern in a Latin square, he will not appreciably bias the estimate of error, as their chance of occurrence is very small (1/672 for a 5 × 5 diagonal square, for example). However, the exclusion of the complementary arrangements by formal application of restricted randomization is possibly worth while, not because it eliminates any general bias over all admitted arrangements, but because it eliminates arrangements that are particularly likely to give an overestimate of error. It should not be forgotten that the experimenter is much more concerned with the trustworthiness of estimates obtained from the arrangement actually selected, than with the behavior of these estimates in a hypothetical population of all admissible arrangements.

It may be asked, if the Knut-Vik squares are known to be on average more accurate, why not always use these, and accept an overestimate of error? There are two objections to this. First, valid estimates of error are in fact often required in experimental work, not only as a basis for tests of significance and fiducial limits in individual experiments, but also for investigating secondary points, for example, variation in treatment effects, over a set of experiments. Second, randomization not only provides valid estimates of error, it also eliminates distortions, which can be large, because, for example, one treatment always occurs in a fixed relation to another. With a randomized design the experimenter can examine his results objectively without continually looking over his shoulder to see if the apparent conclusions require qualification because of some statistical oddity in the design.

4. Estimation and tests of significance

Fisher, I think, tended to lay undue emphasis on the importance of formal tests of significance in experimental work. Many experiments have as their main object the estimation of effects of one kind or another. Often it is well known before the experiment is started that the treatments tested will have some effect. What is then required is an efficient estimate of these effects and a valid and reasonably accurate estimate of their error.

In part this emphasis on tests of significance is attributable to the way in which the subject developed, and to the fact that in the simpler types of experiment the treatment means furnish efficient estimates, whereas the correct estimation of error requires more subtle theory and more extensive computation; in part to the demands of experimenters, particularly biologists, to many of whom the attainment of a significant result seemed more or less equivalent to a new scientific discovery. Fisher himself was also much concerned with the logic of inductive inference, in which tests of significance play a central part.

The emphasis on tests of significance has undoubtedly had unfortunate consequences, both at the practical level, and in theoretical work. Too much effort has been devoted to the investigation of minor points of little real importance. This has resulted in proliferation of alternative methods of analysis, hedged about with restrictions and qualifications, to the confusion of the practical worker.

There is a logical point of some importance concerning alternative tests of significance on the same material. Although with large samples alternative tests which have similar power functions may be expected to give similar results when applied to a given set of data, provided these data conform to the basic assumptions on which the tests are based, this is by no means so with small samples. Consequently two statisticians applying two different tests, both of which are "reasonable," may arrive at very different conclusions. This situation is, to say the least, unfortunate.

An example is provided by the randomization test in experimental design. Fisher originally gave an example of this test in *The Design of Experiments* ([5], 1935) to provide confirmatory evidence of the validity of the t test on the type of data to which it is usually applied without hesitation by the practical statistician. Unfortunately, this was taken to imply that the randomization test, because it made fewer assumptions, was somehow better, and that if the two tests did not agree on a particular set of data, the t test was incorrect. Following this line of thought, Welch [11] evolved a method of "correcting" F tests in randomized block and Latin squares so as to conform approximately to randomization tests.

Fisher did not regard the regular use of randomization and other nonparametric tests as reasonable. As he wrote in the second edition of *The Design of Experiments* ([5], 1937), "they were in no sense put forward to supersede the common and expeditious tests based on the Gaussian theory of errors." He did not, however, ever seriously discuss the question of what should be done when alternative tests give different verdicts, and in various passages in *Statistical Methods for Research Workers* [4], which were never amended, encouraged the statistician to look around for the test giving the highest significance. This is a pity.

5. Fixed, random, and mixed effects models

The analysis of variance, in the form originally proposed by Fisher, and developed by him and his coworkers, rapidly became the accepted method of analyzing replicated experiments. Once the requirements of orthogonality were understood it was successfully applied to very complex types of experiment, for example, those involving a hierarchy of split plots, partial or total confounding and fractional replication.

In addition to providing estimates of error and tests of significance for the various classes of effect, the results of an analysis of variance of experimental data can, if required, be used to estimate the variance components attributable to different classes of effect. Indeed in *Statistical Methods for Research Workers* ([4], 1925) the reader is first introduced to the analysis of variance in this context, as an alternative to intraclass correlation; this, as Fisher said, was "a very great simplification."

For most experiments the estimation of variance components is irrelevant, but such estimates are sometimes required. When, for example, in plant breeding work a random sample of varieties is selected for test, the varietal component of variance may be of interest. Similarly, when a fertilizer is tested on several fields selected at random, the component of variance of the response will represent the true variation in response to the fertilizer (apart from year to year variation).

All this was well known before the war and accepted by those using the analysis of variance to interpret experimental results. Unfortunately, after the war a new concept of fixed and random effect models was introduced. The trouble appears to have started with a paper by Eisenhart [2], which discussed the assumptions underlying the analysis of variance. In the course of this discussion he distinguished what he termed "Model I" or the *fixed effects model* and "Model II" or the *random effects model*. Although there is no real difference in his treatment of these two models, different symbols are used and equivalent formulae for expectations of mean squares in consequence look different. Furthermore, Eisenhart appeared to think that there was a genuine difference between them, or at least encouraged his readers to believe this. He wrote:

"*Which Model—Model I or Model II?* In practical work a question that often arises is: which model is appropriate in the present instance—Model I or Model II? Basically, of course, the answer is clear as soon as a decision is reached on whether the parameters of interest specify *fixed relations*, or *components of random variation.*"

Be that as it may, the hare, once started, could not be stopped. Differences in formulae, arising from differences in definition, soon intruded, and before long it was represented that the tests of significance which could be correctly applied would differ for the two models. (For later developments see a review by Plackett [9]; the discussion on this paper is also worth reading.)

What are the facts? The first and crucial point to recognize is that whether the factor levels are a random selection from some defined set (as might be the case with, say, varieties), or are deliberately chosen by the experimenter, does not affect the logical basis of the formal analysis of variance or the derivation of variance components. Once the selection or choice has been made the levels are known, and the two cases are indistinguishable as far as the actual experiment is concerned. The relevance of the various variance components that can (but need not) be calculated will of course depend on whether the levels can be regarded as approximating to (or are actually) a random selection from some population of levels of interest, but the tests of significance will not be affected.

There is an analogy here with the classical problem of determining the error of a linear regression coefficient, in which, it may be remembered, it was sometimes claimed that allowance had to be made for the fact that the observed x were a sample from some population of x, whereas Fisher rightly insisted they could be taken as known.

The difference in definition, though I have not traced it to its source, appears to be in the interaction constants of an $A \times B$ table ($p \times q$ levels, k replicates). If A and B are regarded as random, the cell (r, s) of the table is taken to have a "true" value of

$$\alpha_r + \beta_s + \gamma_{rs}, \tag{5.1}$$

where α_r, β_s and γ_{rs} are members of populations with variances σ_A^2, σ_B^2 and σ_{AB}^2. The marginal mean of the true values for level r of A will then be

$$\alpha_r + \frac{1}{q} \sum_s (\beta_s) + \frac{1}{q} \sum_s (\gamma_{rs}), \tag{5.2}$$

and the mean square for A in the analysis of variance will have expectation

$$\sigma_e^2 + k\sigma_{AB}^2 + kq\sigma_A^2. \tag{5.3}$$

If B is regarded as fixed, the γ are redefined so as to have zero marginal means over B, that is, to satisfy the conditions

$$\sum_s (\gamma_{rs}) = 0. \tag{5.4}$$

The expectation for the A mean square is then

$$\sigma_e^2 + kq\sigma_A^2. \tag{5.5}$$

This restriction serves no useful purpose. If it is not imposed, the random and fixed models have identical mean square expectations. All that has to be remembered is that the A mean square contains a term in σ_{AB}^2 as well as σ_A^2, and similarly for the B mean square. The factorial case then conforms to the convention customarily adopted for a hierarchical classification, where the mean square for a given level contains variance components for that level and all lower levels.

The above differences in expectations have given rise to the belief that the tests of significance differ for the random and fixed models. Denoting the mean squares by S_A, S_B, S_{AB}, S_E, we can compare S_A with S_E or S_{AB}. The comparison S_A/S_E tests whether for the levels of B in the experiment (whether chosen or obtained by random selection) the effect of A averaged over these levels of B differs from zero. The comparison S_A/S_{AB} tests whether, when the levels of B are a random sample of all possible levels, there is any average effect of A over all possible levels of B.

The latter test is clearly not relevant unless the levels of B can be regarded as a random sample of all levels. Even then I would submit that it is pointless, for if there are interaction terms (σ_{AB}^2 and therefore the γ not zero) their averages over all levels of B will in general not be zero, and thus A effects exist, whereas when there are no interactions S_A/S_E is the appropriate test. The appropriate test for interactions is S_{AB}/S_E, or, more conservatively, if $S_A < S_{AB}$, $[S_{AB}, S_A]/S_E$, where [ ] indicates the combined mean square.

Confidence limits for the population average of the A effects can be obtained directly from S_{AB}, but here another complication intrudes. If A has more than two levels and there is real variation in the A effects over levels of B, there is no reason to expect that S_{AB} will be homogeneous. Thus, if A represents fertilizer levels and B places, the variance of some general measure of response, for example, the linear effect, may be expected to be greater than that of other effect components, for example, the quadratic. Even if A represents varieties the same is likely to be true, for varieties which differ greatly on average may be expected to vary more markedly in their place to place differences. Consequently, for the investigation of place to place differences and similar issues, it is imperative to partition the effects d.f. into single d.f. (usually, but not necessarily, orthogonal) with similar partition of the effects × places d.f. For such investigations to be fruitful fair replication of places is necessary, otherwise the number of effects × places d.f. associated with any particular effect d.f. will be small.

A paper by Harter [7] on the analysis of split plot designs exemplifies the extreme state of confusion that can arise from these differences of definition and from treating replicates as an additional factor R more or less on a par with the treatment factors.

Harter first lists the expectations of the mean squares for A (whole plot factor), B and R, with a, b and r levels, respectively, and their interactions, in terms of σ_A^2, σ_B^2, σ_R^2, σ_{AB}^2, σ_{AR}^2, and so forth, for fixed and random effect models for A and B ("R always regarded as random"). He then gives the list of test ratios shown below, where ρ is the correlation between the subplots in the sense that if σ_w^2 and σ_s^2 are the whole and subplot error variances, additional to σ_{AR}^2 and so forth, $\sigma_w^2 = \{1 - (b-1)\rho\}\sigma^2$ and $\sigma_s^2 = (1-\rho)\sigma^2$.

Test ratios for A. If B is fixed, use S_A/S_{AR} (exact test). If B is random, use S_A/S_{AR} (assumes $\sigma_{AB}^2 = 0$) or S_A/S_{AB} (assumes $\sigma_{AR}^2 = 0$, $\rho = 0$). Satterthwaite test: $S_A/(S_{AR} + S_{AB} - S_{ABR})$; Cochran test: $(S_A + S_{ABR})/(S_{AR} + S_{AB})$.

Test ratios for B. If A is fixed, use S_B/S_{BR} (exact test). If A is random, use S_B/S_{BR} (assumes $\sigma_{AB}^2 = 0$) or S_B/S_{AB} (assumes $\sigma_{BR}^2 = 0$). Satterthwaite test: $S_B/(S_{BR} + S_{AB} - S_{ABR})$; Cochran test: $(S_B + S_{ABR})/(S_{BR} + S_{AB})$.

Test ratio for A × B. Use S_{AB}/S_{ABR} (exact test).

If these were indeed the appropriate tests it might, as he states, be a "crucial" question whether or not S_{BR} and S_{ABR} should be pooled to make up subplot error. Far from being crucial, during the many years I have been responsible for the analysis of split plot experiments I have never considered the two components worth separation. The only reason for separation would be to examine, from the results of many experiments, whether there was evidence for variation in response to the subplot factor over replicates; for this purpose S_{BR} might be tested against S_{ABR}. This question of differential response has been fairly thoroughly examined in connection with confounded experiments [8], [12], [15] because in such experiments differential response would appear as an apparent interaction between other factors. All the evidence indicates that it is of negligible magnitude in agricultural field trials, even with fertilizer treatments.

Suffice to say that, of the five experiments quoted by Harter, S_{BR} is *less* than S_{ABR} in three, and in none is the difference significant at the 5 per cent level. Yet, following Bozivich, Bancroft, and Hartley [1], he recommends separation in three experiments, in one of which S_{BR} is less than S_{ABR}!

Incidentally, it may be noted that Harter's arguments have really little to do with split plots. The same partition of error d.f. can be made in an ordinary $A \times B$ factorial experiment, and the same confused situation would arise if his arguments were valid.

6. Nonfactorial response surface designs

Nonfactorial response surface designs, of which rotatable designs are an example, were originally introduced to determine the optimum levels of several factors in industrial processes. They have occasionally been advocated for use in preference to ordinary factorial designs in agricultural field trials whose primary object is to determine optimal levels of fertilizer components. This use seems to me to be very questionable, for the circumstances are very different.

An example of a design that recently came to my notice may bring out the basic objections to rotatable designs. This design was in fact more extreme than a rotatable design, but will serve as an illustration. The design was for a set of trials on nitrogen and phosphate with treatments as in figure 3.

The experimental results were sent to Rothamsted because analysis on a desk calculator was too onerous. It was onerous for us too, as the required formulae had to be worked out, and some special programming was required. This is a very real disadvantage of these designs.

In correspondence on this design we made the following comments. It would have been much better to use the conventional 3×3 design, starting each fertilizer at zero level with increments of 30 lb/acre. The information on what happens at the corners of the ordinary conventional 3×3 design is of great importance, and in the present design the curvatures are ill determined because of nonorthogonality. The point is illustrated by table I of variances in three replicates of (I) a 3×3 factorial design with factors 5 or 35 or 65 lb N per acre, and 0 or 30 or 60 lb P_2O_5 per acre, and (II) the design adopted (table I).

TABLE I

COMPARISON OF VARIANCES

	I	II
Linear	1/72	1/36
Linear × Linear	1/192	1/12
Quadratic	1/96	5/192

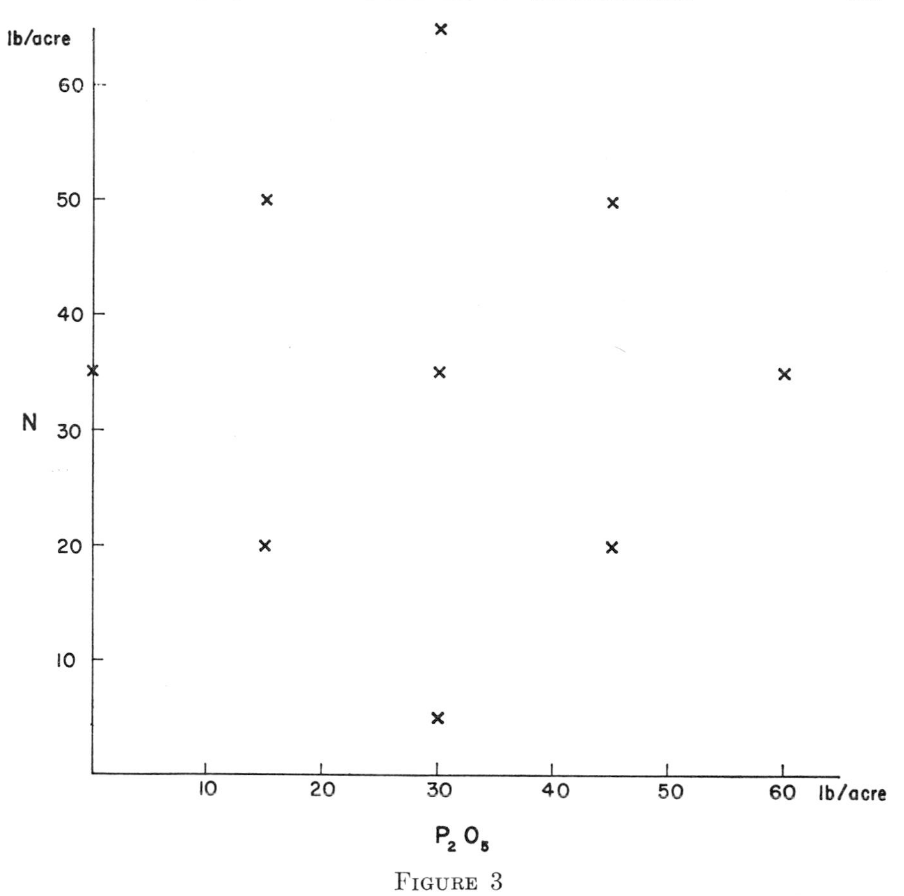

FIGURE 3

Response surface design for N and P.

Of course, Box's designs have advantages over factorial designs. They enable a given number of factors to be tested on fewer units (plots) if necessary, they are particularly suitable for sequential experimentation (indeed they were first introduced for this purpose) and in certain circumstances they are less sensitive to departures from the assumed form of response surface. These points may well be relevant in industrial experimentation, particularly when errors are small, so that loss of efficiency becomes less important, but none is a major consideration in present day fertilizer experiments.

The much lower accuracy of the design is, of course, mainly due to the fact that there are fewer points at the extremes of the permissible ranges of N and P. But for fertilizer components, and I suspect for many factors in other types of experiment, there are no grounds for reducing the range of one factor at the extremes of the others. Exploration of a rectangular area of the response surface, rather than a circular area, is more appropriate. If it were certain that the factors interacted positively, as often happens with fertilizer components, there might be a case for omitting the 0, 2 and 2, 0 combinations in a 3×3 design, but negligible interactions are quite common, and in some sets of experiments the existence of strong negative interactions has been established.

The lowest level of a factor need not, of course, be zero. When, for example, it is known from previous work that good responses to N are to be expected, levels of N_1, N_2, N_3 might be adopted for the factorial scheme. It is then useful to include some additional plots (for example, one in each block of a confounded $3 \times 3 \times 3$ factorial) with, for example, no N and the intermediate levels of P and K. This indeed is now common practice, and our computer program for the analysis of $3 \times 3 \times 3$ factorials provides for such additional treatments, and also for summarizing groups of experiments.

One further general point. A single experiment on fertilizers, as indeed is true in much other experimental work, is not expected to give final answers on all points. We do not ask that the optimal dressings should be determinable from one experiment. To do so would demand an impracticably large experiment, and would in any case be relevant only to the chosen field in the current season. All that we expect is reasonably accurate information on the general responses to the separate factors. Information on the curvatures of the response curves and interactions between components, which is required for the calculation of optimal dressings, is gradually accumulated as the experimental program proceeds. This is another fundamental difference between the industrial and agricultural situation.

7. Computers and experimental design

Electronic computers are radically altering the computational problems associated with the analysis of experiments, and this has some influence on design. When only desk calculators were available, it was imperative that the arithmetical computations should be kept simple. Now we can, potentially, face much more extensive numerical work if this results in compensating advantages.

I say "potentially" advisedly, for it is unfortunately true that as yet very few computers are programmed to provide full analysis of the more complex types of experiment that are in current use. It may be objected that the need for this is not great, as such experiments were in the past satisfactorily analyzed on desk calculators, and this can continue. This is only partially true, as anyone who has contact with extensive experimental programs knows. Backlogs of unanalyzed results build up, and multivariate techniques involving covariance analysis are seldom used; these are frequently required for experiments in which many variates are observed. Missing observations also cause much trouble.

We have shown at Rothamsted that powerful general programs for the analysis of wide groups of designs can be written. We have, for example, recently written a general factorial program which handles designs with partial (balanced or unbalanced) or complete confounding (including $3^p \times 2^q$ confounded designs), split plots (including successive splits), and even, although this was not planned, fractional replication. Covariance, missing values, and preliminary processing of the data are included, and the results are presented in a form acceptable to those familiar with desk calculators.

I consider that an urgent task facing statistical departments in universities and research institutes is to provide programs of this type on the computers to which they have access. This need not be an onerous task if a cooperative effort is made, and programs are written in a common language such as Fortran or Algol. It would indeed be a great help to practical experimenters if there were a set of standard statistical programs common to all computers of the requisite size. But such standard programs must be good ones. Many of the programs at present available are insufficiently general and unsatisfactory statistically.

To return to the question of how computers will influence design, it is clear that designs requiring the inversion of matrices can now be faced. This is already leading to the development of designs on mixtures (simplex designs); I believe much further interesting work lies ahead of us here. But I would emphasize that in general additional computational complications should not be introduced unless they can be shown to have compensating advantages, and that in many circumstances computationally simple designs, because of their balanced properties, are in fact the most efficient.

Further work for which I believe computers will be particularly useful is the combined analysis of sets of experiments, and the analysis of the accumulated results of long term experiments. The techniques required for these tasks require much further development to which I hope we at Rothamsted will make a useful contribution.

REFERENCES

[1] H. Bozivich, T. A. Bancroft, and H. O. Hartley, "Power of analysis of variance test procedures for certain incompletely specified models. I," *Ann. Math. Statist.*, Vol. 27 (1956), pp. 1017–1043.

[2] C. Eisenhart, "The assumptions underlying the analysis of variance," *Biometrics*, Vol. 3 (1947), pp. 1–21.

[3] R. A. Fisher, "The arrangement of field experiments," *J. Ministry Agric.*, Vol. 33 (1926), pp. 503–513.

[4] ———, *Statistical Methods for Research Workers*, Edinburgh, Oliver and Boyd, 1925–1958.

[5] ———, *The Design of Experiments*, Edinburgh, Oliver and Boyd, 1935–1960.

[6] P. M. Grundy and M. J. R. Healy, "Restricted randomization and quasi-Latin squares," *J. Roy. Statist. Soc. Ser. B*, Vol. 12 (1950), pp. 286–291.

[7] H. L. Harter, "On the analysis of split-plot experiments," *Biometrics*, Vol. 17 (1961), pp. 144–149.

[8] O. Kempthorne, "A note on differential responses in blocks," *J. Agric. Sci. Camb.*, Vol. 37 (1947), pp. 245–248.

[9] R. L. Plackett, "Models in the analysis of variance," *J. Roy. Statist. Soc. Ser. B*, Vol. 22 (1960), pp. 195–217.

[10] O. TEDIN, "The influence of systematic plot arrangement upon the estimate of error in field experiments," *J. Agric. Sci. Camb.*, Vol. 21 (1931), pp. 191–208.

[11] B. L. WELCH, "On the z-test in randomized blocks and Latin squares," *Biometrika*, Vol. 29 (1937), pp. 21–52.

[12] F. YATES, "Complex experiments," *J. Roy. Statist. Soc.*, Suppl. 2 (1935), pp. 181–247.

[13] ———, "A further note on the arrangement of variety trials: quasi-Latin squares," *Ann. Eugen.*, Vol. 7 (1937), pp. 319–332.

[14] ———, "Sir Ronald Fisher and the design of experiments," *Biometrics*, Vol. 20 (1964), pp. 307–321.

[15] F. YATES, S. LIPTON, P. SINHA, and K. P. DAS GUPTA, "An exploratory analysis of a large set of $3 \times 3 \times 3$ fertiliser trials in India," *Emp. J. Exp. Agric.*, Vol. 27 (1959), pp. 263-275.

Mathematical equivalence of the various models

Distinction must be made between models which are mathematically equivalent in the sense that the estimates of the parameters appertaining to either model can be derived from the estimates obtained from the other and models for which this does not hold. The regression models

$$Y = \alpha + \beta x + \gamma x^2 + \delta x^3$$

and

$$Y = \mathbf{A} + \mathbf{B}\xi_1 + \mathbf{C}\xi_2 + \mathbf{D}\xi_3$$

for example, where ξ_1, ξ_2, ξ_3 are orthogonal polynomials of degree 1, 2, 3, are mathematically equivalent. The estimates a, b, c, d of α, β, γ, δ can be derived from the estimates A, B, C, D of **A, B, C, D,** and vice versa. The orthogonal polynomial form has the practical advantage that solutions for first and second degree polynomials are also directly available; moreover inversion of a matrix is not required.

Apart from the fact that in certain circumstances negative variance components are admissible in the random model, the fixed and random models are mathematically equivalent. It is not to be expected, therefore, that the choice of one model rather than another will contribute materially to the interpretation of the results, provided always, of course, that the meaning of the various parameters is properly understood.

Admissibility of negative variance components

I should have mentioned in my paper that in certain circumstances negative variance components are admissible in the random model. This applies not only to cross classifications but also to hierarchical classifications. The only important example I have come across is that of sampling from finite populations. If the plots of an agricultural experiment are sampled for yield, say, by dividing each plot into h sampling units of which k are taken at random, and the ordinary hierarchical model without restrictions is used, the yield of sampling unit r of the plot with treatment p in block q of a randomized block experiment can be expressed as

$$y_{pqr} = \mu + \alpha_p + \beta_q + \gamma_{pq} + \delta_{pqr}$$

the γ being normal deviates with variance σ_E^2, and the δ normal deviates with variance σ_S^2, without restrictions. The mean square expectations when the mean squares are expressed in units of the sampling unit are then

	M.S.	Expectation
Experimental error	E	$k\sigma_E^2 + \sigma_S^2$
Sampling error	S	σ_S^2

In plot units the experimental error, therefore, has expectation $\sigma_E^2 + \sigma_S^2/k$. If all h units had been taken the expectation would have been $\sigma_E^2 + \sigma_S^2/h$.

Cases can arise, e.g. when there is intense competition between plants on a plot, in which the experimental error when the whole plot is sampled is less than σ_S^2/h. This can be allowed for by admitting negative estimates of σ_E^2 (minimum value, $-\sigma_S^2/h$), but is more elegantly dealt with by redefining the δ for each plot as deviations from the mean of h normal deviates with variance σ_S^2, with a compensating increase in the variance of the γ from σ_E^2 to σ'^2_E, such that $\sigma'^2_E = \sigma_E^2 + \sigma_S^2/h$. A variant of this latter specification was used in my investigation of the efficiency of sampling cereal experimental plots for yield (Yates and Zacopanay, 1935).*

With this specification the expectation of the experimental error mean square in plot units is

$$\sigma'^2_E + \frac{1}{k}\sigma_S^2\left(1 - \frac{k}{h}\right)$$

This reduces to σ'^2_E when all h sampling units are taken, i.e. the 'fixed' model expectation, and to $\sigma_E^2 + \sigma_S^2/k$ when only a very small proportion of sampling units is taken, i.e. the 'random' model expectation. $1 - k/h$ is the well-known correction for finite sampling.

For the estimation of the efficiency of sampling it is immaterial which specification is chosen. If the first is taken the efficiency (proportion of information retained when k instead of all h units of a plot are taken) is expressible as

$$\left(\sigma_E^2 + \frac{1}{h}\sigma_S^2\right)\bigg/\left(\sigma_E^2 + \frac{1}{k}\sigma_S^2\right)$$

if the second, as

$$\sigma'^2_E\bigg/\left[\sigma'^2_E + \frac{1}{h}\sigma_S^2\left(1 - \frac{k}{h}\right)\right]$$

Equating the mean squares E and S to their expectations in terms of σ_S^2 and σ_E^2 or σ'^2_E and substituting in the above formulae, we obtain in either case the expression

$$1 - \left(1 - \frac{k}{h}\right)\frac{S}{E}$$

for the estimate of the efficiency of sampling, with a minimum value, imposed by the lower bounds, of 0.

* J. A. Nelder (*Biometrika*, **41**, 544-548, 1954) discussed the similar situation in split-plot experiments. He there gives a formal mathematical interpretation of negative components of variance.

Example of the various models in two-factor experiments

Consider the factors: levels of nitrogen (N), levels of potash (K), varieties (V), locations (L). Taking N and K as fixed and V and L as random, and two-factor experiments (p and q levels respectively, r replicates) on N and K, N and V, and L and V respectively, the mean squares have the expectations shown in Table 1.

Table 1. Degrees of freedom and mean square expectations for various models

	D.F.	N and K (both fixed)	N (fixed) and V (random)	L and V (both random)
N or L	$p - 1$	$\sigma_E^2 + qr\sigma'^2_N$	$\sigma_E^2 + r\sigma_{NV}^2 + qr\sigma_N^2$	$\sigma_E^2 + r\sigma_{LV}^2 + qr\sigma_L^2$
K or V	$q - 1$	$\sigma_E^2 + pr\sigma'^2_K$	$\sigma_E^2 + pr\sigma'^2_V$	$\sigma_E^2 + r\sigma_{LV}^2 + pr\sigma_V^2$
Int'n	$(p - 1)(q - 1)$	$\sigma_E^2 + r\sigma_{NK}^2$	$\sigma_E^2 + r\sigma_{NV}^2$	$\sigma_E^2 + r\sigma_{LV}^2$
Error		σ_E^2	σ_E^2	σ_E^2

It has been claimed that this presentation is helpful in indicating which tests of significance are relevant in different circumstances. The main point that it brings out is that the effect of a factor, whether random or fixed, which is crossed with a factor which is fixed, can be tested against error without qualification, whereas for a factor which is crossed with a factor which is random, there are two appropriate tests: one against experimental error, which is relevant to the effects of the factor over the actual selection of levels of the random factor, and the other against interaction, which is relevant to the whole population of levels of the random factor. However, when making the latter test, as is pointed out in Paper XII, we cannot assume that the interaction degrees of freedom are homogeneous, and consequently the degrees of freedom for N must be partitioned into single degrees of freedom, with a corresponding partition of $N \times V$.

There therefore seems to be no point in writing down the variance components given by the conventional fixed and random models merely to determine which tests of significance are appropriate (or, more important, which components of error should be used to estimate confidence limits). Certainly the erroneous recommendations by Harter in the paper quoted can in the main be attributed to this procedure.

Variance components

The component of variance due to levels of a quantitative factor, whether or not it is crossed with other factors, is clearly dependent on the range of levels included. The concept of variance components is consequently of no value for such factors, and we need not consider further the fixed model exemplified by the $N \times K$ experiment.

The variance components of qualitative factors such as varieties require further study. Take first the experiment on locations and varieties. It might at first be thought that σ_V^2 was the varietal component of variance. What is usually required, however, is the varietal component of variance within locations. Clearly if the experiment had been conducted at a single location, only this would have been available: if more than one location is included, whether randomly selected or hopefully 'representative', the average of the within locations components for the different locations can be taken. This is equal to $\sigma_V^2 + \sigma_{LV}^2$, as is easily seen from the specification of the random model. It can be obtained directly by pooling the V and $L \times V$ terms in the analysis of variance, i.e. by setting out the analysis in hierarchical form with the term 'varieties within locations'. If on the other hand we require only the component common to all locations, of which those selected can be considered a random or at least a representative sample, this is given by σ_V^2.

The difference between blocks and locations should be noted. In an agricultural experiment blocks might be regarded as different locations within a field. In general, however, it is reasonable to attribute variations in the differences between the varieties in the different blocks to fertility differences, etc., which have nothing to do with varieties, i.e. to experimental error.

The situation in the N and V experiment is somewhat different. A substantial $N \times V$ interaction implies that the varietal component of variance is likely to differ considerably for different levels of N. This is not necessarily so, but is to be expected because a large interaction term is likely to be due to a large linear component of differential response for varieties; if so, the varietal variance will be greater at one extreme of N than the other, or at both extremes compared with the centre of the range.

We therefore have to consider whether we are content with the average varietal component of variance over all included levels of N, as with locations and varieties, or whether the component at some particular level of N is required. The former is obtained by pooling the N and $N \times V$ terms. The latter cannot be obtained from the analysis of variance, even if the N and $N \times V$ terms are split into orthogonal N components, as these components cannot be assumed to be uncorrelated: it is most easily obtained by direct calculation from the varietal means for the required level of N. What should not be taken as the estimate is σ'^2_V. This is always less than the average over all levels by an amount $\sigma^2_{NV}(p-1)/p$.

The real distinction is not between factors whose levels are randomly selected and those in which the levels are arbitrarily chosen, but between factors for which the interaction components in the model can be specified not too unreasonably as random uncorrelated values with the same variance—uncorrelated, that is, before restrictions are imposed—and factors for which this assumption is patently false. Quantitative factors such as fertilizer levels belong to the latter category, as is easily seen by considering actual examples. For the former category the random model is always appropriate. For the latter category a more complex model involving the components of the response curve, e.g. linear and quadratic components, and interactions of these components, is required, and is indeed habitually used by those concerned with the analysis of this type of experiment.

A model of this type is 'fixed' in the sense that the interaction components for any given level of a factor crossed with a quantitative factor have zero sums, but differs radically from a model in which the interaction components for the whole of the two-way table are taken to be random variates with the same variance and uncorrelated except for the correlations introduced by the restriction to zero sums. If the more realistic model is adopted for the N and V experiment, for example, with orthogonal components N_1, N_2, . . . for N, the mean square expectations for the V, N_1V, N_2V, . . . terms in the analysis of variance will be $\sigma^2_E + pr\sigma'^2_V$, $\sigma^2_E + \lambda_1 r\sigma^2_{N1V}$, $\sigma^2_E + \lambda_2 r\sigma^2_{N2V}$, . . ., where the values of λ_1, λ_2, . . . depend on the conventions regarding units. As before, the variance component of varieties averaged over all levels of N can be derived from the pooled sum of squares for V and all the $N \times V$ terms, and equals $\sigma'^2_V + \lambda_1\sigma^2_{N1V}/p + \lambda_2\sigma^2_{N2V}/p + \ldots$

A genetical example

The main use of variance components is in genetics. As an example we may take the estimation of heritability from sib analysis. If the same sires are used to fertilize dams at several locations by artificial insemination, locations and sires form a cross-classification. With l locations, s sires, d dams at each location, and p progeny of each dam, the degrees of freedom and mean square expectations for measurements on the progenies, using the random model, are those of Table 2.

Table 2. Degrees of freedom and mean square expectations in a breeding experiment

	D.F.	Mean square expectation
Locations	$l - 1$	$\sigma_W^2 + r\sigma_D^2 + rd\sigma_{LS}^2 + rds\sigma_L^2$
Sires	$s - 1$	$\sigma_W^2 + r\sigma_D^2 + rd\sigma_{LS}^2 + rdl\sigma_S^2$
Locations × sires	$(l - 1)(s - 1)$	$\sigma_W^2 + r\sigma_D^2 + rd\sigma_{LS}^2$
Dams within sires	$ls(d - 1)$	$\sigma_W^2 + r\sigma_D^2$
Within progenies	$lsd(p - 1)$	σ_W^2
	$lspd - 1$	

If locations are taken as fixed the expectation of the sires mean square becomes $\sigma_W^2 + r\sigma_D^2 + rdl\sigma'^2_S$.

Some geneticists, I am told, are in doubt whether locations should be treated as fixed or random and are concerned about this because different values are obtained for heritability. This can only be because of a misconception of the meaning of σ_S^2 and σ'^2_S. For calculating heritability the components of variance within locations are usually what is required; for dams only that within locations (and within sires) is available. As explained above, the variance for sires within locations is given by $\sigma_S^2 + {}_{LS}^2$, not by σ_S^2 or σ'^2_S $(= \sigma_S^2 + \sigma_{LS}^2/l)$, and can be obtained as a single component by a completely hierarchical analysis of variance, combining sires and sires × locations. In the orthogonal case the separation of the sires × locations interaction is preferable, as this will reveal any differential selective advantage of sires at different locations. Moreover it serves to indicate that the sires within locations component is less accurate than might be presumed from the pooled degrees of freedom.

1969 FRANK YATES

BIBLIOGRAPHY

Books, monographs, etc.

Notes on the application of the method of least squares to the adjustment of triangulation and level and traverse networks (with J. Clendinning). Gold Coast: Government Printing Office, 1929.

Gold Coast Survey: Tables for use in the Department. Gold Coast: Government Printing Office, 1929.

Report on the survey and adjustment of a network of precise traverses measured between 1924 and 1930 in the southern portion of the colony (with H. Wace). Gold Coast: Government Printing Office, 1931.

The design and analysis of factorial experiments. Commonwealth Bureau of Soil Science, Technical Communication No. 35, 1937.

Statistical tables for biological, agricultural and medical research (with R.A. Fisher). Oliver and Boyd, Edinburgh. 1st Edition 1938, 2nd Edition 1942, 3rd Edition 1948, 4th Edition 1953, 5th Edition 1957, 6th Edition 1963.

Sampling methods for censuses and surveys. Griffin, London. 1st Edition 1949, 2nd Edition 1953, 3rd Edition 1960.

Journal papers

1932 The effect of climatic variations on the plasticity of soil (with G.W. Scott Blair). *J. agric. Sci.*, **22**, 639–646.

1933 On the validity of Fisher's *z* test when applied to an actual example of non-normal data (with T. Eden). *J. agric. Sci.*, **23**, 6–17.

1933 The principles of orthogonality and confounding in replicated experiments. *J. agric. Sci.*, **23**, 108–145.

1933 The analysis of replicated experiments when the field results are incomplete. *Emp. J. exp. Agric.*, **1**, 129–142.

1933 An experiment on the incidence and spread of angular leaf-spot disease of cotton in Uganda (with C.G. Hansford, H.R. Hosking and R.H. Stoughton). *Ann. appl. Biol.*, **20**, 404–420.

1933 The formation of Latin squares for use in field experiments. *Emp. J. exp. Agric.*, **1**, 235–244.

1934 The analysis of multiple classifications with unequal numbers in the different classes. *J. Amer. statist. Ass.*, **29**, 51–66.

1934 Observer's bias in sampling observations on wheat (with D.J. Watson). *Emp. J. exp. Agric.*, **2**, 174–177.

1934 A complex pig-feeding experiment. *J. agric. Sci.*, **24**, 511–531.
An interesting early example of a factorial experiment on pigs, using individual feeding.

1934 Contingency tables involving small numbers and the χ^2 test. *Suppl. J.R. statist. Soc.*, **1**, 217–235.

1934 The 6 × 6 Latin squares (with R.A. Fisher). *Proc. Camb. phil. Soc.*, **30**, 492–507.
An enumeration of the 6 × 6 Latin squares, and a proof of Euler's classic supposition that there is no 6 × 6 Graeco-Latin square. All 6 × 6 squares are shown to be derivable from 12 standard types by permutation and interchange of rows, columns and letters.

1935 Some examples of biased sampling. *Ann. Eugen.*, **6**, 202–213.

1935 Complex experiments. *Suppl. J.R. statist Soc.*, **2**, 181–247.

1935 The estimation of the efficiency of sampling, with special reference to sampling for yield in cereal experiments (with I. Zacopanay). *J. agric. Sci.*, **25**, 545–577.
The efficiency of sampling plots of cereal experiments for yield is examined, using data from eighteen experiments harvested by sampling methods to assess the average loss of information. The paper contains two topics of general interest, the interpretation of the analysis of variance in sampling problems, and the elimination of bias from estimates each of which is a ratio of two mean squares.

1935 The place of quantitative measurements on plant growth in agricultural meteorology and crop forecasting. Conference of Empire Meteorologists, *Meteorological Office Publication No. 393*, 169–172.

1936 Incomplete Latin squares. *J. agric. Sci.*, **26**, 301–315.
The least squares solution is given for Latin squares in which a row, column, or treatment, or a row and column, or either and a treatment is missing.

1936 Crop estimation and forecasting: indications of the sampling observations on wheat. *J. Minist. Agric.*, **43**, 156–162.

1936 A new method of arranging variety trials involving a large number of varieties. *J. agric. Sci.*, **26**, 424–455.

1936 Incomplete randomized blocks. *Ann. Eugen.*, **7**, 121–140.

1936 Applications of the sampling technique to crop estimation and forecasting. Manchester Statistical Society.

1937 A further note on the arrangement of variety trials: Quasi-Latin squares. *Ann. Eugen.*, **7**, 319–332.
An extension of Paper vi to the elimination of fertility differences appertaining to the rows and columns of a square of plots.

1938 The analysis of groups of experiments (with W.G. Cochran). *J. agric. Sci.*, **28**, 556–580.

1938 The gain in efficiency resulting from the use of balanced designs. *Suppl. J.R. statist. Soc.*, **5**, 70–74.
An early example of an actual balanced incomplete block design used in an experiment on the feeding of dietary supplements to school children.

1938 Orthogonal functions and tests of significance in the analysis of variance. *Suppl. J.R. statist. Soc.*, **5**, 177–180.

1939 The comparative advantages of systematic and randomized arrangements in the design of agricultural and biological experiments. *Biometrika*, **30**, 440–469.

A paper by Barbacki and Fisher (*Ann. Eugen.*, **7**, 189–193, 1936) attacking the half-drill strip method gave rise to a regrettable controversy between Gossett ('Student') and Fisher in which Gossett justifiably claimed that Barbacki and Fisher's criticism was wrongly based. Here I endeavour to put the record straight, and show that the weaknesses of the half-drill strip method are very different from those stated by Barbacki and Fisher. The paper also contains some interesting points about randomization in other types of design.

1939 The recovery of inter-block information in variety trials arranged in three-dimensional lattices. *Ann. Eugen.*, **9**, 136–156.

A companion paper to Paper viii.

1939 Factors influencing the percentage of nitrogen in the barley grain of Hoosfield (with D.J. Watson). *J. agric. Sci.*, **29**, 452–458.

1939 The analysis of Latin squares when two or more rows, columns or treatments are missing (with R.W. Hale). *Suppl. J.R. statist. Soc.*, **6**, 67–79.

The least squares solution for this problem, with an appendix on the effect of restrictions on the fitted parameters.

1939 Tests of significance of the differences between regression coefficients derived from two sets of correlated variates. *Proc. Roy. Soc. Edin.*, **59**, 184–194.

1939 The adjustment of the weights of compound index numbers based on inaccurate data. *J.R. statist. Soc.*, **102**, 285–288.

1939 An apparent inconsistency arising from tests of significance based on fiducial distributions of unknown parameters. *Proc. Camb. phil. Soc.*, **35**, 579–591.

1940 Lattice squares. *J. agric. Sci.*, **30**, 672–687.

Another companion paper to Paper viii, on the recovery of interblock information in quasi-Latin squares.

1940 The recovery of inter-block information in balanced incomplete block designs. *Ann. Eugen.*, **10**, 317–325.

1940 Modern experimental design and its function in plant selection. *Emp. J. exp. Agric.*, **8**, 223–230.

1941 Fertilizer policy in war-time. The fertilizer requirements of arable crops (with E.M. Crowther). *Emp. J. exp. Agric.*, **9**, 77–97.

An important paper, illustrating how the results from a heterogeneous collection of experiments – in this case on fertilizer responses – can be combined to give a coherent picture.

1942 Statistical problems in field sampling for wireworms (with D.J. Finney). *Ann. appl. Biol.*, **29**, 156–167.

1942 Influence of changes in level of feeding on milk production (with D.A. Boyd and G.H.N. Pettit). *J. agric. Sci.*, **32**, 428–456.

A re-analysis and re-interpretation of a large set of excellent Danish experiments and other related experiments on the feeding of dairy cows – also relevant to wartime policy.

1943 Methods and purposes of agricultural surveys. *J.R. Soc. Arts*, **91**, 367–379.

1944 The manuring of farm crops : some results of a survey of fertilizer practice in England (with D.A. Boyd and I. Mathison). *Emp. J. exp. Agric.*, **12**, 163–176.

1946 A review of recent statistical developments in sampling and sampling surveys. *J.R. statist. Soc.*, **109**, 12–43.

1946 Agriculture and the food crisis. *Contact*, **2**, 93–94.

1946 The place of statistics in agricultural research. *Agric. Progr.*, **21**, 1–11.

1947 The analysis of data from all possible reciprocal crosses between a set of parental lines. *Heredity*, **1**, 287–301.

1947 The analysis of contingency tables with groupings based on quantitative characters. *Biometrika*, **35**, 176–181.

1947 Technique of the analysis of variance. *Nature*, **160**, 472.

1947 The influence of agricultural research statistics on the development of sampling theory. *Proc. Int. Stat. Cong.*, **3**, 27–39.

1948 Systematic sampling. *Phil. Trans. R. Soc.* **241**, 345–377.

1948 Operational research. *Nature*, **161**, 609.

1949 The design of rotation experiments. Proceedings of the Commonwealth Agricultural Bureaux Conference on Tropical and Sub-Tropical Soils, Commonwealth Bureau of Soil Science Technical Communication **46**, 142–155.

An attempt, not very successful, I fear, to introduce order into the design of rotation experiments, which because of the many variants that can be introduced defy orderly classification. There is, however, a useful discussion on the possibilities of designs of strictly limited duration.

1949 The relative yields of different crops in terms of food and their responses to fertilizers (with D.A. Boyd). *Agric. Progr.*, **24**, 1–11.

1949 The residual manurial value of feeding stuffs consumed on grass (with D.A. Boyd, E.M. Crowther and J.R. Moffatt). *J.R. agric. Soc.*, **110**, 104–114.

1950 The place of experimental investigations in the planning of resource utilization. *Proceedings of the United Nations Scientific Conference on the Conservation and Utilization of Resources.*, **1**, 192–196.

1950 Agriculture, sampling and operational research. *Bull. Inst. Int. Statist.*, **32**, 220–227.

1950 Experimental techniques in plant improvement. *Biometrics*, **6**, 200–207.

1950 The place of statistics in the study of growth and form. *Proc. Roy. Soc. B.*, **137**, 479–488.

1951 Bases logiques de la planification des experiences. *Ann. Inst. H. Poincaré*, **12**, 97–112.

1951 Quelques développements modernes dans la planification des expériences. *Ann. Inst. H. Poincaré*, **12**, 113–130.

1951 Manuring for higher yields. "Four thousand million mouths". Ed. F. le Gros Clark and N.W. Pirie, pp. 44–73. Oxford University Press.

1951 The influence of *Statistical Methods for Research Workers* on the development of the science of statistics. *J. Amer. statist. Ass.*, **46**, 19–34.

1951 The Survey of Fertilizer Practice: An example of operational research in agriculture (with D.A. Boyd). *Brit. Agric. Bull.*, **4**, 206–209.

1951 Crop prediction in England. *Bull. Inst. Int. Statist.*, **33**, 295–312.

1951 Statistical methods in anthropology (with M.J.R. Healy). *Nature*, **168**, 1116.

1952 Analysis of a rotation experiment. *Bragantia*, **12**, 213–236.

1952 Principles governing the amount of experimentation required in developmental work. *Nature*, **170**, 138–140.

1952 George Udny Yule: 1871–1951. *Obit. Not. Roy. Soc.*, **8**, 309–323.

1953 The use of fertilizers on food grains (with D.J. Finney and V.G. Panse). *Indian Council of Agricultural Research, Research Series*, No. 1.

1953 Selection without replacement from within strata with probability proportional to size (with P.M. Grundy). *J.R. statist. Soc. B*, **15**, 253–261.

1953 The wider aspects of statistics. *J. Ind. Soc. Agric. Stat.*, **5**, 109–118.

1954 The analysis of experiments containing different crop rotations. *Biometrics*, **10**, 324–346.

1955 The use of transformations and maximum likelihood in the analysis of quantal experiments involving two treatments. *Biometrika*, **42**, 382–403.

Experiments for which the results are percentages (e.g. of diseased plants) can be analysed by first transforming the data. For transformations such as the logit transformation, a maximum likelihood solution is required. An iterative procedure similar to that used for probit analysis is described for experiments with two treatments. Now that computers are available the same procedure can be used generally.

1955 A note on the application of the combination of probabilities test to a set of 2×2 tables. *Biometrika*, **42**, 404–411.

1955 Report to the Government of India on Statistics in Agricultural Research (with D.J. Finney). *F.A.O. Report No.* 358.

1957 An automatic programming routine for the Elliott 401 (with S. Lipton). *J. Assoc. Computing Machinery*, **4**, 151–156.

1957 Routine analysis of replicated experiments on an electronic computer (with M.J.R. Healy and S. Lipton). *J.R. statist. Soc. B.*, **19**, 234–254.

1958 A note on the six-course rotation experiments at Rothamsted and Woburn (with H.D. Patterson). *J. agric. Sci.*, **50**, 102–109.

1958 The use of an electronic computer in research statistics: four years experience (with D.H. Rees). *Comp. J.*, **1**, 49–58.

1958 Second report to the Government of India on Statistics in Agricultural Research. *F.A.O. Report No.* 987.

1959 An exploratory analysis of a large set of 3 × 3 × 3 fertilizer trials in India (with S. Lipton, P. Sinha and K.P. Das Gupta). *Emp. J. exp. Agric.*, **27**, 263–275.
A combined analysis of 238 3 × 3 × 3 single replicate fertilizer trials on rice, using a computer. The analysis illustrates the partition of main effects and interactions into orthogonal contrasts, as in Paper v, with a hierarchical analysis of variance for each contrast.

1959 Wilfred Leslie Stevens. *J.R. statist. Soc. A*, **122**, 403–404.

1960 A general programme for the analysis of surveys (with H.R. Simpson). *Comp. J.*, **3**, 136–140.

1960 The use of electronic computers in the analysis of replicated experiments, and groups of experiments of the same design. *Bull. Inst. Agron. Gembloux*, **1**, 201–211.

1961 Marginal percentages in multiway tables of quantal data with disproportionate frequencies. *Biometrics*, **17**, 1–9.

1961 The analysis of surveys: processing and printing of the basic tables (with H.R. Simpson). *Comp. J.*, **4**, 20–24.

1962 Electronic computation and data processing for research statistics (with M.J.R. Healy). *Bull. Inst. Int. Statist.*, **39**, 305–318.

1962 Computers in research – promise and performance (Presidential Address). *Comp. J.*, **4**, 273–279.

1962 Sir Ronald Aylmer Fisher, F.R.S. *Nature*, **195**, 1151–1152.

1962 Sir Ronald Aylmer Fisher: 1890–1962. *Rev. Inst. Int. Statist.*, **30**, 280–282.

1962 Sir Ronald Aylmer Fisher. *Biometrics*, **18**, 437–454.

1963 A specialised autocode for the analysis of replicated experiments. (with J.C. Gower and H.R. Simpson). *Comp. J.*, **5**, 313–319.

1963 Fisher's contributions to the design and analysis of experiments. *J.R. statist. Soc. A.*, **126**, 168–169.

1963 Ronald Aylmer Fisher: 1890–1962 (with K. Mather). *Biogr. Mem. roy. Soc.*, **9**, 91–129.

1963 What is wrong with the teaching of statistics? "Contributions to Statistics", pp. 485–494. Oxford: Pergamon Press.

1964 An example of the analysis of uniformity trial data on an electronic computer (with A.J. Vernon and S.W. Nelson). *Emp. J. exp. Agric.*, **32**, 25–30.

1964 Sir Ronald Fisher and the Design of Experiments. *Biometrics*, **20**, 307–321.
This traces the way Fisher developed the design and analysis of experiments. It is of interest that the development of methods of analysis preceded the improvement of design.

1964 Fiducial probability, recognisable sub-sets and Behrens' test. *Biometrics*, **20**, 343–360.

1964 How should we reform the teaching of statistics ? (with M.J.R. Healy). *J.R. statist. Soc. A*, **127**, 199–210.

1964 Two decades of Surveys of Fertilizer Practice (with D.A. Boyd). *Outlook*, **4**, 203–210.

1965 Net shyness. Appendix to "Observations on noctule bats (*Nyctalus noctula*) captured while feeding" by the Earl of Cranbrook and H. Barrett. *Proc. zool. Soc. Lond.*, **144**, 22–23.

1965 A fresh look at the basic principles of the design and analysis of experiments. *Proc. Fifth Berkeley Symposium on Mathematical Statistics and Probability*, **4**, 777–790.

1965 Discussion of reports on cloud seeding experiments. *Proc. Fifth Berkeley Symposium on Mathematical Statistics and Probability*, **5**, 395–397.

1966 A general computer programme for the analysis of factorial experiments (with A.J.B. Anderson). *Biometrics*, **22**, 503–524.

This program was an important advance in organizing the analysis of replicated experiments of varied types on computers. An appendix on least square solutions in experimental design explains certain points not previously dealt with.

1966 Computers, the second revolution in statistics (First Fisher Memorial Lecture). *Biometrics*, **22**, 233–251.

1968 Theory and practice in statistics (Presidential Address). *J.R. statist. Soc. A.*, **131**, 463–477.

1969 The evolution of a survey analysis program. "New Developments in Survey Sampling" pp. 59–80. New York: Wiley–Interscience.

INDEX

Additivity (see also Interaction), 81, 100
Analysis of covariance, 154, 279
Analysis of variance, 5, 17, 32, 45, 123, 160, 203, 213, 234, 237, 242, 268, 272

Balance, 97, 235
Balanced incomplete blocks, 72, 109, 183 f., 207
Balanced pairs, 107, 185
Bias, 44, 60, 92, 140, 260, 269
Block size, 102, 107

Comparisons, contrasts, 34, 85, 151, 245
Computers, 278
Confounding, 3 f., 79, 93, 184, 279
Controls, 151, 153
Costs and losses, 258
Cycle, 232

Diagonal squares, 60,270
Diallel cross, 213 f.
Differential responses, 109, 261, 275
Dummy treatments, 33, 97

Eden, T., 63
Efficiency, 73, 91, 102, 152, 178, 189
Efficiency factor, 160, 162, 190, 203
Estimation, 137, 271

F distribution, see Z distribution
Factorial experiments (see also Confounding), 71 f., 279
Factorial experiments, 2^n, 32, 79, 93, 269
Factorial experiments, 3^n, 95, 278
Factorial experiments, $2^p \times 3^q$, 96, 279
Fisher, R.A., 3, 45, 75, 78, 267 f.
Fitting constants, see Least squares
Fixed and random effects models, 273

Graeco-Latin and higher squares, 101, 194

Half-drill strip method, 107
Harter, H.L., 275

Interaction, 6, 55, 80, 83, 93, 123, 274
Inter-block comparisons, 93, 203 f.

Knight's move (Knut-Vik) squares, 60, 270

Latin squares, 19, 59 f., 90, 101, 103, 163, 183
Latin squares, missing value in, 46
Lattice designs, 149 f., 207
Least squares, 6, 9, 24, 45, 156, 215, 244

Mathematical equivalence of models, 281
Missing values, 43 f., 129, 171, 279

Normal equations, see Least squares

Orthogonality, 3 f., 43, 82, 235, 243 272

Partial confounding, 95, 279
Phase, 232
Plant-breeding, 213 f.
Pooling estimates of error, 109, 124
Pseudo-factorial designs, see Lattice designs

Quasi-factorial designs, see Lattice designs
Quasi-Latin squares, 269

Randomization, 59, 122, 267, 272
Randomized blocks, 3, 43, 100, 103, 183
Randomized blocks, missing value in, 46
Regression (see also Analysis of covariance), 129, 274
Response curves and surfaces, 86, 258, 275, 276
Rotatable designs, 276
Rotation experiments, 231 f.

Semi-Latin squares, 92
Semi-weighted mean, 138
Sequence, 231
Series of experiments, 83, 121 f., 209, 278, 279
Series (in rotation experiments), 233
Significance tests, 7, 49, 271, 274
Split-plot experiments, 16, 87, 275, 279
Strip experiments, 18
Technical experimentation, 121, 257 f.,

Tedin, O., 60
Transformation sets (of Latin squares), 64

Variance components, 273, 284
Variety trials, 149

Weather, 132, 261
Weighing designs, 100
Weighting, 136, 165, 204
Welch, H.L., 272
Wishart, J., 4, 44, 102

Z distribution, 63

Worked examples

3×2, non-orthogonal, 8
3×2^4, Latin square with row and column treatments, 21
3^3 in blocks of 9, quality × quantity, 35
2^3 with 9 missing values, 48
3×4, split-plot, 88
2^3 in blocks of 4, 94
3^3 in blocks of 9, single replicate, 98
Variety trials, 6 places × 2 years, 126
3^3 series, 15 places, 141
Systematic controls, 170
7×7 lattice, 171
$4 \times 4 \times 4$ lattice, 176
5 treatments, balanced pairs, 186
7 treatments, balanced blocks of 4, 197
9 treatments, balanced blocks of 4, 207
21 varieties, balanced blocks of 5, 208
12×12 diallel cross, 223